Physics

Against The Odds

By

Pharis E. Williams

Williams Research

http://www.physicsandbeyond.com/

Library of Congress Cataloging-in-Publication Data

Williams, Pharis E.

  Physics, Against the Odds : physics, nuclear physics, cosmology, gauge theory / Pharis E. Williams

Includes bibliographical references and index.

ISBN 978-0-615-26722-7

1. Physics, 2. Nuclear Physics, 3. Particle Physics, 4. Field Theory, 5. Gravity, 6. Cosmology, 7. Dark Matter, 8. Dark Energy, 9. Waves

Cover design by Debbie Balog (see www.dbalog.com)

Send correspondence to the author of this book, see

www.physicsandbeyond.com.

ii

To Jeri, without her love, support and urgings this work would not be

CONTENTS

Forward

Throughout my years of education, I heard teachers and professors state "This is what we now know" when presenting the latest findings of science or the latest theoretical formulation. As the years went on, I was amused by the fact that such pronouncements were later found to be in error when new experimental observations or new theoretical constructs overturned the outdated concepts. However, the scientific community never seemed to miss a beat. Those old "What we thought we knew" concepts were just replaced by the new "What we now know" proclamations.

It has been said by others that science has two kinds of practitioners – those who are "Keepers of the Flame" and those who truly exercise original thought and challenge the sanctuary of "what we now know." The "Keepers" are very important to science in that they maintain the base of knowledge and transmit it to following generations of scientists. These worthies add knowledge incrementally through extensions of current theory or through more detailed and precise experiment. I certainly do not mean to belittle these folks. They are the backbone of science and without their contributions there would not be a science to carry forward. In my humble estimation, the "Keepers" make up almost all of the scientific community. We are all "Keepers" to a certain extent.

However, the earthshaking new directions in science have come from the very few who are the original thinkers. The most obvious member of this august group is Albert Einstein. His Special and General Relativity Theories brought about a revolution in scientific thought that resonates even today. Others – I would call them "Keepers" – applied and extended his theories, but his thoughts remain the basis of all current physics. Such scientists are very rare indeed. I firmly believe that I have had the great pleasure of associating with a member of this rare breed.

I met Pharis E. Williams in 1973 at the U. S. Naval Postgraduate School in Monterey, California. We were classmates in the Nuclear Physics curriculum and all 18 "Nuke" students had been assigned study offices in room 008 in the basement of the Physics building. While I was studying for upcoming classes, a fellow walked into my cubicle sporting the kind of southern twang that belonged in a farmer's field somewhere. Willie, as we learned to call him, was already holding forth on the fundamental basis of physics before we had even started our upper level course work on the subject. Our classmates and I quickly learned that this Missouri farm-boy possessed a world class intellect and a very inquisitive mind.

Willie learned quickly that I would take any opportunity to lay aside my studies to hear him pontificate at the white board. I was fascinated by his questioning approach to "what we now know" of physics. He wanted to start from the foundational assumptions of physics and ask "Why?" – Why those assumptions and not others? Had we – the physics community – gone down the wrong paths by making ever more "simplifying" assumptions? Were we ignoring contrary experimental evidence in order to maintain our cherished theories? I was captured by his imagination immediately. I objected to the oft repeated admonition that "We are not allowed to ask that question" when dealing with certain areas of advanced physics and I suspected that "what we now know" had serious flaws. Willie was a kindred soul who had better math skills than I and I wanted to hear more.

As we advanced through the physics curriculum, Willie was able to put deeper thought and more mathematical rigor into his developing theory and I was able to follow most of his thoughts and some of his math. As the theory took on more structure and more of our classmates became convinced we were in the presence of genius, we started to campaign for a symposium to display what Willie was thinking. We wanted to have Willie put his work before a gathering of the math and science faculty. As his reputation grew throughout the Postgraduate School, they finally agreed and a symposium was arranged. When the day arrived, I sat at the back of the meeting to observe the reactions of the assembled professors as Willie displayed his command of the broad expanse of physics and math. As the briefing started I noticed that many in the audience appeared to be disdainful of this countrified junior naval officer who had such radical ideas about the fundamental basis of physics. However, as Willie laid out his logical thought progression using all the tools of advanced mathematics, including tensor analysis, the attitudes changed perceptibly. That day, Willie made many converts among the Math, Science, and Computer Science faculties. They were not ready to accept his challenge to the fundamentals of science, but they were convinced he possessed the ability and the tools to cause many to think afresh about "What we now know." His classmates were convinced we were in the presence of a true "Original Thinker." I remain convinced to this day.

Upon graduation we each went our separate ways as service in the Navy required. I had adopted the view that sailors belong on ships and ships belong at sea. I was a sailor and, therefore, I reported to my new ship and continued my career as a naval officer. However, I never forgot those days of listening to Willie's theory and tilting against the windmills of science. Throughout the intervening years wherever we were, Willie kept me informed about the newest developments in the theory until that day he called to ask if I would consider coming to Los Alamos, New Mexico, to assist with his research. With the help of a friendly admiral, we arranged for me to be assigned to Los Alamos National Laboratory (LANL) as a Military Research Associate and I set my naval career aside for a life of research.

My first duties at LANL were to reacquaint myself with the mathematics of the Dynamic Theory such that I could properly support its claims to others. In doing this I started with the three fundamental assumptions – the conservation of energy, Caratheódory's statement of the Second Law of Thermodynamics (generalized to mechanical entropy or you can't get somethin' for nothin'), and a rule for comparing one system's mechanical entropy to another's. If we allow the need for a fifth dimension (and that seems to be the case) these three laws lead inexorably to an expansion of Maxwell's well known equations of electromagnetism. Once I had convinced myself of that fact, I approached several scientists at LANL and other research organizations to seek their response to such a result. What I found was disheartening. Some responded that "Mister Williams' theory is too simplistic." Others said that "Mister Williams' theory is too complicated." I even stumbled across scientists who were still smarting from Willie's uncovering of the shortcomings of the Nuclear Laboratories in determining the yield of certain weapons and the resultant dangers posed by that revelation. I was thrown out of several offices. I have been thrown out of better places. There were also positive responses. Gian Carlo Rota, a well known and respected mathematician from Princeton University and a Fellow at LANL, remarked that "Mister Williams does not make mathematical errors" after he reviewed the entire theory.

Ignoring the naysayers as suffering from "not invented here" syndrome or possibly "I've spent my whole career climbing this tree and now you say I'm in the wrong tree?" I took my findings on the road to try and garner funding for research. We were able to stir up quite a lot of interest, but no principal.

One such foray was a briefing I gave to the Chief of Naval Operations (CNO) Executive Panel – the staff responsible for new technical initiatives. After several weeks of exposing the staff to Willie's theory, the Director approached me and asked if he could borrow my viewgraphs. I agreed and he promised to tell me what he could about their use. A few days later I received a call from him saying that I should watch TV that night and I would understand the impact of my briefing. That night President Reagan was on all channels with his Strategic Defense Initiative speech, later called the "Star Wars" program. The Director of the Executive Panel called later and said that my viewgraph showing the expanded Maxwell Equations was on the President's desk as he made the speech. As I returned to Los Alamos, I was floating on the euphoria of hoped for research funding from such a successful trip.

When I returned to the LANL DoD Program office, where Willie and I worked, the corridor was abuzz with the prospect of billions of dollars in funding that would be made available to conduct the innovative research required to meet the President's vision. Our protestations that the Dynamic Theory was one element of the President's calculations had no effect on the thinking of the DoD Program staff. They were busy dusting off every unfunded wild and semi-wild idea they had put forth for years and could not be bothered with an unproven theory. LANL scientists were not being irresponsible or un-American. They were just captive to the "what we now know" philosophy and trying to put flesh on the President's vision with science and technology that they understood and agreed with. After all, Los Alamos got its start with the development of nuclear weapons and that was one area they understood well. Use of those types of devices and countering them were natural research for the LANL scientific staff.

In the years since, Willie has made several attempts to publish his theory in various journals with some limited success. However, the physics establishment has yet to pay serious attention to the inescapable logic and solid mathematical basis for the theory. This book is a long envisioned attempt to publish a layman's summary of the theory so that more people are exposed to the thinking that Pharis *Willie* Williams has done over the years. The Dynamic Theory gores too many sacred oxen of current physics to be considered by the "Keepers." It also clearly indicates an inductive coupling among electric, magnetic, and gravity fields, something that needs serious investigation. Read this book with an open mind and the mindset of an "Original Thinker." I hope you will be stimulated to think further.

James O. (Oke) Shannon, CDR, USN, retired

Preface

I knew before I began that my chances of succeeding started at zero and went up to mighty slim. So why did I embark on an effort to determine if the currently accepted physics could be found to all come from some more fundamental basis? I have read somewhere that close to 90 percent of all scientists who ever lived were alive now. Most of these were above average intelligence. Many of these scientists were geniuses. What gave me the audacity to think that I might succeed where so many others could not?

I was approaching it my way.

I don't recall ever receiving an award for the most outstanding person at following orders. Perhaps, in a way I am still a two year old.

I think I had another advantage. I had nothing to lose.

I was just starting a postgraduate education and I had switched my major from electrical engineering to physics. If I dug into the bowels of physics where all the fundamental concepts, assumptions and laws were kept I would surely learn more physics than if I just went to class and learned what they had to tell me. If I learned that I could not find a more fundamental set of laws from which the currently accepted ones could be derived I still would have learned a lot of physics. Perhaps I would have learned more than if I had just gone along.

But how might my way be any different from the hundreds, perhaps thousands, of others? I had the log across the road. This is a concept, an approach to life, or a philosophy that I had picked up as a boy and had used as a method of troubleshooting malfunctioning electrical circuits and life in general.

I was probably only 8 or 9 years old when I overheard a local southern Missouri gentleman (then called a hillbilly) answer a non-local's (then called a city slicker) request for directions in the following manner. The old gentleman told the non-local to follow a road down a hollow for a few miles and when he came to a fork in the road to take the right hand fork. He pointed out that the right hand road was the wrong road, but told him to take it anyway. The old gentleman then said that a couple of miles further there would be a log across the road. That log was confirmation that the right hand fork was indeed the wrong road. The non-local was to go back to the fork in the road, take the left hand fork, and then he couldn't miss what he sought.

For many years I thought this was just a procedure the old gentleman used to reinforce the fact that the non-local should take the left hand fork first. Then I begin to see that there was something to be learned in knowing why one road was right and another was wrong. For example, how many times in life have we come up against a difficulty that was really an indication that we were on a wrong road and we simply cut the log up and proceeded on down the wrong road? At what point do we ask ourselves whether or not a difficulty is a log across our path that indicates we should go back the way we had come and try a different route?

The log across the road constituted my way.

I looked into the currently known physics looking for logs across the road of progress using the various theories currently accepted. I found that all had logs except one. I traveled down each of these roads to learn why the wrong ones were wrong so that I might know when I arrived at current physics while I traveled on down the one road without a log across it.

The many hours, days and years I have spent in this effort has given me a different appreciation of just how slim the odds were that I could succeed.

I also have a very different appreciation for how trying a man's obsession can be for his wife and how appreciative I should be for the toleration I have been given.

Chapter 1 Why Physics?

For thirty some years I have been thinking about physics, writing about physics and obsessed with physics. For thirty some years I have listened as learned scientists tell me I am wrong. However, in that time only one individual has pointed to a mistake in the mathematical logic I have used and this turned out to answer a major question. For thirty some years I wished that I could have found a job as a theoretical physicist. Now I thank God that I never found such a job. If I had found employment as a theoretical physicist I would have had to go along with the current thinking in physics in order to keep such a job. There have been a few close friends and associates who have maintained a steady interest in my research and have given me their support as they could. Still it has been a long, lonely road. Why did I start down this road? Why did I stay on it?

Nuclear Weapons Safety

My early job in the US Navy was an enlisted electrician's mate. In this job I changed light bulbs, fixed motors, ran the generators and basically attended to all things associated with electrical power. After going to sea, attending some service schools and advancing in rank, I was selected to attend college to obtain a bachelor's degree in electrical engineering. I was sent to the University of Colorado at Boulder, CO for four years including the three summer sessions. This gave me time enough to seek two bachelor's degrees, one in electrical engineering that the Navy sent me to college to pursue and one in applied math that I wanted. I had always felt that math was the language one needed to know in order to understand science and this was my opportunity to learn extra math.

Upon graduating from college I was assigned to the Navy's Officer Candidate training and left there as an ensign in the naval officer corps. I was then sent to a World War II class diesel submarine where I learned very quickly that I did not want to remain under water for those long periods of time. I then served on a destroyer escort as gunnery and missile officer where I either did a good job and my commanding officer recommended me to the admiral or the commanding officer wanted to get me off his ship and thought the admiral would help him achieve that goal. Either way I was assigned to the flotilla admiral's staff as the missile officer. After that tour of duty I was sent to another admiral's staff.

Too much staff duty restricted a young officer's ability to acquire the tickets he needed for advancement so I sought a way of breaking out of the staff assignments. I found what I thought I could use when the Chief of Naval Operations issued an invitation for young officers to volunteer for duty in Vietnam. I wasn't so hot to go to war as I was to get off the staff duty that this would provide. So I went to war. At least until my oldest son was hit by a car, actually he ran into the car, and was hospitalized. Then I was sent home a little early.

One of the incentives to get young officers to volunteer to go in-country Vietnam was to guarantee them their choice of postgraduate school when they returned home. However, in my case I did not want to go the Miami University in Ohio because my son could not be transported for a while. So I took a billet at the Nuclear Weapons Training Group, Pacific located on North Island Naval Air Station. There I was assigned as the Course Supervisor for some of the weapons safety, handling and employment courses.

A short time after my arrival my immediate supervisor, Commander Philip Anthony, noticed that I had more than the average number of math credits in my

2

college transcript and came to me with an additional request. For some time he had felt that the design of nuclear weapons did not meet the required safety design criteria but he did not have the educational tools that would allow him to properly investigate the potential problem. Would I do it for him?

I agreed, but had I known what it would lead to I might have not been so agreeable. I began a project to learn what the design criteria were, why they were established and how the current weapons matched up to these requirements. In the course of this project I obtained some experimental data from the Sandia National Laboratory and the Naval Ships Laboratory in Carderock, MD plus some computer calculations from the Lawrence Livermore National Laboratory. The combined results from these inputs, plus my own calculations, showed that some weapon design change that had already been projected and halted needed to be implemented and other weapons' design needed additional changes. Of course these change requirements were not met with much enthusiasm. As a matter of fact, they caused more than one senior naval officer to threaten my career. The only thing that saved my career was the fact that, despite what the national labs first claimed, they discovered that I was correct and the changes not only were needed, but were implemented.

The final meeting, at which a decision on the action to be taken was to be made, was held after I had orders to the Naval Postgraduate School in Monterey, CA. Fresh upon the heels of that meeting in which I had taken on a room full of PhDs from the national laboratories and had prevailed in the face of their spirited opposition, I had the audacity to think I could trust my mental capabilities in the realm of physics. So I changed my degree choice from electrical engineering to physics.

My Way

When I arrived at the Naval Postgraduate School I began going to the library and reading all the physics books I could find that talked about the fundamentals, or foundations, of physics. I also talked with several physics professors asking them questions about the foundations of physics and if they could be done differently.

A torpedo without an instruction book

Then, after a few weeks the chairman of the physics department called me into his office. He talked rather slower than he usually did and seemed to be choosing his words very carefully. He likened me to a navy torpedoman who has just been given a new torpedo without an instruction book. I was going to tweak the knobs without knowing what would happen and it would blow up in my face. He told me I really should listen to what they had to say about physics before I told them how I thought physics ought to be.

I had given this notion a great deal of thought before I started my searching through the foundations of physics and I saw two ways of going. One way, the way the chairman of the department suggesting, was the way almost all students followed. This included beginning from scratch and going through the standard curriculum which starts with Newtonian physics and progresses through special relativity, quantum mechanics to general relativity following virtually the history of physics theoretical development. I felt there was a disadvantage in taking this course of study in that by the time I finally learned all the currently available knowledge of physics I would be so inundated with the now-known that new thoughts would be severely hampered. I had just seen what sort of trouble that taking the old master's word and not questioning it brought to the nuclear weapons community. I did not wish to be guilty of blindly following where others had gone.

On the other hand the alternative seemed to be worse. What I saw as the alternative was to try to develop a philosophy of science before I absorbed the philosophy that comes along with the current view of physics as taught. This approach is asking for trouble. It is, in a way, the height of arrogance and naïveté to think that one individual could find a better way to look at physics than all of the past great scientists. I offered myself the fall-back position that by studying the foundations of physics I would learn more physics this way using this motivation than I would following the usual path.

It might be helpful to say a little about why I felt something was wrong in physics that needed to be fixed in order to lend a measure of understanding to why I felt so strongly that I needed to find my own way as I also learned what is known.

Though I had often asked "Why?" when confronted with some new assumption or adopted postulate, the first really puzzling facet of current physics I encountered was the concept of relativistic kinetic energy from Einstein's special theory of relativity. The puzzling part was that it depended upon the speed of light independent of the mechanism by which this energy might be transferred. To better illustrate what puzzled me, consider the transfer of energy between two charged particles on collision courses. If the particles have near-miss trajectories, then the energy is primarily transferred by the electrical forces between the charges. From the view of retarded potentials, or the concept of a limiting speed of electromagnetic signal transmission, it is rather easy to accept the energy transferred being dependent upon this limiting velocity. But suppose the particles are uncharged and the interaction is strictly a gravitational one. Again the concept of a limiting signal speed would imply that the energy exchanged between the particles depend upon this limiting velocity. But is it the same as the limiting signal velocity for the electromagnetic case? Do gravitational waves travel at the same speed as electromagnetic waves?

Einstein, in the special theory of relativity, adopted the position that the constancy of the speed of light forces a modification of Newton's dynamic law. This modification implies that all forces have the same limiting velocity, namely, the speed of light. There exists an abundance of theoretical and experimental evidence that the speed of light becomes the limiting velocity whenever electromagnetic forces are involved. The point that bothered me was whether other forces, such as gravitational, should also have the same limiting velocity. Though we have had reports of the detection of gravitational waves, we have no experimental determination of the speed of a gravitational wave. Therefore, I objected to the viewpoint that the modification to Newton's law should be applied to all forces without some additional justification.

Let me describe an analogy which may not hold in the strictest sense yet may serve to illustrate my point of view. A river, flowing toward the sea, carries energy with it. The speed with which this energy can move from one point to another is the velocity of the river's current. The river produces a force on a boat tied up to a pier on the river. When the boat is set adrift, this force accelerates the boat. However, the maximum velocity to which the river can accelerate the boat is the current velocity; this is the velocity with which the energy of the river can propagate.

From this point of view the speed of light, being the propagation velocity of electromagnetic energy must be the limiting velocity associated with electromagnetic forces. Certainly nature would be much simpler if all forces have the same limiting velocity. Yet without some experimental evidence of the propagation of gravitational energy, I found it difficult to feel comfortable with Einstein's modification of Newton's law justified by electromagnetic experimental evidence and arguments of simplicity.

The fundamental philosophical viewpoint that the force depends upon velocity and vanishes as the velocity approaches the limiting velocity raises another question concerning Einstein's modification of classical mechanics. Under Einstein's modification Hamilton's principle is written with a relativistic mass which depends upon the velocity and a velocity independent force. Does this represent a different philosophy or are both views equivalent? More specifically, are the "real" concepts to be taken as a mass independent of velocity together with a velocity dependent force or should we associate the velocity dependent relativistic mass and velocity independent forces with the "real" world? Or does it make any difference which we chose?

Quantum mechanics posed problems for me also. Not that I didn't believe that nature worked with bunches, but that I didn't believe quantization was the most fundamental of all theories. Nature has a lot of phenomena that involve quantization. For example, all stringed musical instruments make use of quantization of string motion to make the beautiful tones they produce. Perhaps we all learned as children that should we tie one end of a rope to a tree and wiggled the other end in an up-down motion we could cause standing waves in the rope. Electromagnetic wave guides have boundary conditions that establish standing waves within the guides. These phenomena were all cases where some restriction, or boundary condition, created the quantization. That argued to me that quantization was not the fundamental element, but was caused by some restriction to the system.

The fact that time seemed to always go in one direction while the equations of motion physics gave us allowed time to go in both directions did not seem right. What was missing in the foundations of physics that we hadn't yet learned? Thermodynamics had an approach to an arrow of time that we all have learned about from the bull in a china shop or the fact that heat always flows from a hotter body to a colder body.

My efforts to explain my reasons, desires, and yes, my obsession, for physics my way did not seem to sway the department chairman. However, he did inform me that he would be my thesis advisor. I found after graduation that he did not want to be my advisor, but had no choice in the matter. As he explained at a dinner after my graduation, he and the other physicists in the department did not feel they could answer my questions about fundamental physics. Therefore, as chairman of the department it fell to him to take the job because he didn't feel any of the other professors could keep me in check.

Thermodynamics does it

Then the work began. I continued to spend a lot of time in the library. I took all of the math courses I could get the professors to teach. In order to get a course taught enough students needed to enroll to make up a class. I consider myself to be indebted to several classmates as they repeatedly signed up for higher level math and physics courses they didn't need, or want, in order to make a full class for some course I wanted to take.

History records the advancements in physics which came from the efforts of people new to the field. Therefore my lack of training in physics might be turned into an advantage if I sought to determine a philosophical basis unhampered by the directed philosophy that comes from a study of physics as currently taught. This is, as mentioned above, in contradistinction with current practices and procedures of academicism where mastery of current theories generally precedes the development of a new one. To deliberately choose this deviation risks accusations of arrogance and naïveté. On the other hand such a choice seemed the best way of avoiding the danger of becoming so familiar with current ways of thinking as to make it improbable of giving due attention to other ways.

Having decided to look for a new foundation for physics I was faced with the question of how to begin. I recalled some Ozark hill philosophy I overheard as a youngster. A native Ozarkian was giving directions to a stranger who was trying to find a certain fishing hole. The directions went something like this: "See yonder road going down that holler? Well, go down thar 'bout five mile and you'll come to a fork in the road. Take the right hand fork. Now that's the wrong one but you take it anyways. After you've gone a piece, you'll come to a log across the road. Now you know you're on the wrong road. So go back and take the left hand fork. You can't miss it."

A quick review of physics reveals that there are different branches with different sets of fundamental laws or postulates. Though it is easy to see how the distinction between these branches came about, it was difficult for me to believe that nature shared the same divisions. I felt that all natural phenomena should be explained by a single set of fundamental laws. This belief is somewhat like a grove of redwood trees or bamboo forest. Above the ground each tree appears as a distinct plant. Yet we know that below the ground they may be found to grow from the same root system. Thus, I felt that a more fundamental approach might display the unity in nature and that prior attempts at unification in the search for a unified field theory could be likened to attempts to tie the trees together at the tree top level rather than down at the root level.

Is nature symmetrical in time? Does everything run backward in time as well as forward? Obviously, not every process in nature will run backwards, yet the equations of motion in Newtonian and relativistic mechanics are time symmetrical. I believe in an asymmetrical nature and this belief played a role in the eventual selection of fundamental laws.

How then did I use this philosophy to determine a set of generalized laws on which to base an attempt to construct a new approach to physics? Newtonian mechanics fails to describe events involving high velocities, relativistic mechanics fails to describe the atom, and gravitational effects have resisted quantization. If these are viewed as logs and the Ozarkian's directions are followed, then we must retrace our steps and seek another approach rather than attempting to chop up the log and continue to push forward up one of these roads.

The branch of thermodynamics, however, does not appear to have a log somewhere along the way. Thus the thermodynamic laws appeared to be the fork in the road where a new route might be chosen.

However, in mechanics we talk of equations of motion, field equations, and geometry while in thermodynamics we speak of equations of state and equilibrium. If a generalization of the classical thermodynamic laws is adopted, how might we obtain the equations with which we are familiar in mechanics? More particularly, how could this type of general law yield geometry and a variational principle? The second law of thermodynamics can produce a variational principle through principles such as increasing entropy and minimizing free energy, but can it also produce geometry?

This seemed to be a crucial point. If the laws could not produce a geometry, then a geometry would have to be assumed, thus necessitating an additional assumption. The belief that a simple fundamental set of laws should lead to the fundamental principles of the different branches of physics made the thought of additional assumptions abhorrent. The notion that the adopted laws should specify the type of geometry that must be used seemed very satisfying. Newton found that the absolute nature of

10

Euclidean geometry brought undesirable features. Einstein, in his general theory, displayed the benefits that might be gained by going to a more general geometry. He showed that physical phenomena might be displayed as elements determined by certain physical laws. This is essentially the question here. Can a set of laws, which are generalizations of the classical thermodynamic laws, determine the metric elements and hence the geometry?

At this point I seemed to have become stumped. I did not know enough thermodynamics. In my undergraduate education I had nearly flunked thermodynamics. Like most electrical engineering students I didn't think thermodynamics was that important. Now I needed to learn what I had not learned before.

Thermodynamics is really misnamed. As it is taught is should be called thermostatics for we teach primarily about equilibrium states; not the dynamics of going from one state to another, or going from a non-equilibrium state to an equilibrium state. Further, the statements of the second law did not lend themselves to generalization. I had trouble thinking of how to mathematically operate with a second law statement that said something like "Heat always flows from a hotter body to a colder body." I had just taken another thermodynamics course and asked the professor if there existed a more general statement of the second law? The professor sent me to look at Caratheódory's statement of the second law.

Caratheódory had produced a very general statement of the second law that was stated in terms a mathematician might use. It probably came from the fact that Caratheódory was a mathematician.

If one is seeking equations whose solutions tell the motion of a particle or system he needs to have two things. First, he needs to know the underlying geometry. The second thing one needs to arrive at equations of motion is some statement that tells how things are to proceed.

Usually this statement is one that says to look for the minimum distance between two points. Thus, since one has the geometry that describes the distances between two points and then knows he needs to look for the path that gives the minimum distance between the two points he can find how things will go. This is why pilots fly in a great circle arc when traveling between two cities that are great distances apart. The spherical geometry of the Earth means that the minimum distance between two points on the Earth's surface is along an arc called a great circle arc.

By appealing to the mathematics of functions of more than one variable we find that a quadratic form becomes involved when a maximum or minimum is sought. Further, this quadratic form generates a natural geometry for that function. In thermodynamics the stability conditions provide a similar quadratic form and therefore the quadratic form which specifies the stability conditions should form a natural geometry for a physical system governed by laws such as the thermodynamic laws.

Thus the foundations of the theory have been outlined, namely the belief that all physical phenomena should be derivable from a single set of physical laws which are generalizations of the classical thermodynamic laws. Such a theory should be capable of describing all the dynamic events in nature. Therefore it seems appropriate to call it the "Dynamic Theory". Obviously, for such a theory to be tenable it must reproduce, or be consistent with, the various fundamental postulates and/or laws currently used in the various branches of physics. Indeed it should do even more. It should also reduce the number of necessary assumptions and provide an unprecedented unification of physics. Further, there is the possibility that the theory might produce an experimentally verifiable prediction.

The first requirement that should be placed upon the Dynamic Theory is that it reproduces, or be consistent with, current theories. Though a theory which has the capability of displaying a unification of physical theories might have significant value based solely upon this capability, it would become more attractive if it could explain phenomena for which no explanation exists or make some new prediction which might lead to an experimental test of the theory. Since restrictions were placed upon the system in order to show how current theories may be obtained, the easiest way to see the expanded coverage of the theory is to relax one or more of the restrictions and consider a more general system.

Isentropic mechanics

Once I learned about the second law as stated by Caratheódory I began to investigate the potential of this statement to provide a description of mechanical motion. I spun my wheels a lot. I was playing in a new play ground with a whole playground full of new play equipment and I didn't know how to use any of it. Perhaps the most important thing coming from the research contained in my thesis was the potential importance of the role of entropy in mechanical motion. The second important finding was that the second law of thermodynamics required a limiting velocity that was independent of the type of force being considered. The third thing revealed by the research was the difference between thinking of a velocity dependent mass and a velocity dependent force.

No more weapons

Not long before I was to receive orders for a duty station following my graduation from postgraduate school I received a call from the Navy Captain who was the program manager of the weapons modification program started as the result of my prior work on weapons safety. He wished to know if I would be interested in working with him to insure that proper modifications were completed. I told him that, of course, I would be interested. A few days later he called again. This time he indicated that there was more than one admiral in the pentagon that still held hard feelings about the "trouble" I had created with my earlier work. They opposed my working on the project. They didn't even want me to know what was happening in the program. The Captain was powerless to change their minds. He told me the admirals wanted to send me somewhere that would keep me out of the way so they were sending me to teach at the US Naval Academy. What was I supposed to teach? I was being sent to teach theoretical thermodynamics!

5D at the Naval Academy

I learned that I was to teach three sections of theoretical thermodynamics at the US Naval Academy to junior students in the mechanical engineering department, plus one class in engineering orientation to the freshman class. Now I would really have to learn thermodynamics in order to teach it in a class that not only taught the theoretical basis for the naval steam plants but also taught the steam plant operation. This meant I would have to learn how to apply the theoretical thermodynamics to a practical system.

Is mass conserved?

I taught the fall semester without any diversions. But the spring semester had barely begun when a question came to mind. The question: "Was mass really conserved?" This question arose while preparing for my class involving the first law of thermodynamics.

The first law of thermodynamics is a very general law that equates the amount of heat transferred between a system and its surroundings with the change in the system's energy that is not due to the work of any force and the energy due to all the different forces at work. For a steam plant this means there are four work terms that must be considered. These four work terms are the thermodynamic force, the pressure, times the change in volume it causes and the three mechanical work terms that primarily involve friction due to steam flowing through pipes and the work of raising the weight of water up against the force of gravity.

This meant that a classroom presentation of the first law involves writing an equation on the board showing this equality. This equation would look like

$$ dQ = dU - P\,dV - F_x\,dx - F_y\,dy - F_z\,dz \,. $$

The left hand side of this equation indicates the change in heat. The little horizontal bar in the letter d is the mathematical way of saying that this is a small change that depends upon the path taken. Another way of saying this is that it indicates a path dependent differential. The important point with respect to the question concerning the conservation of mass is seen on the right hand side of the equation. There are five independent differentials on the right hand side, dU, dV, dx, dy, and dz. Now the students know that a system of equations with five independent variables must have five equations in order to allow us to find a solution to the system of equations. For our thermodynamic system we can use the equation of state that relates the pressure to the volume, or other variables. Usually the equation of state that is used is the ideal gas law that states the product of the pressure and volume is a constant. That gives us one equation. Next we can use the three Newtonian force laws that specify any loss of pressure due to friction in the pipes or work against the force of gravity. That gives us four equations. We need one more to have a solvable system of equations. The usual comment by the teacher at this point is to state that we know we can always write the mass density as a function of space and time and since mass density in inversely proportional to the specific volume we have the required fifth equation.

But how do we know we can always write mass density as a function of space and time? Of course if mass is conserved we can for the conservation of mass is an equation relating how mass density may vary in space and time. But are we assured that mass is really conserved? We have a very vivid display of the conversion of mass into energy in every atomic bomb. We are also told that Einstein's famous equation

$$E = mc^2$$

means that conservation of mass and energy are related. How can we then question whether or not mass is really conserved?

We will find that when we tied energy conservation to mass conservation using Einstein's relation we have made a very common error of theoretical interpretation. That will become clearer as we go on. For now let us return to our question and ask how we may prove that we can always write mass density as a function of space and time? After failing to find a proof on my own I talked with physicists across campus and on other campuses. Yet all I obtained were statements that did not have the strength of a mathematical proof, but did express the common belief that it must be possible to prove. Many used arguments that included relativity theory similar to what is implied above, but these tend to be circular arguments and are virtually the same as assuming the answer in order to prove your answer.

Finally I talked with a mathematician at MIT and her response was to say that if I did not feel comfortable with the proofs that were offered an alternative route could be taken. Her suggestion was to assume that mass was independent of space and time and see if the deductions that this assumption produced were supported by experiment. If there existed experimental evidence that a deduction based upon the assumption that mass was independent of space and time differed from data collected

16

from experiment then the assumption must be wrong and mass must be dependent upon space and time. This is a common route taken by mathematicians when one approach to a proof seems too difficult.

So what are some of the deductions that come from assuming mass is independent of space and time? There are many. They are varied and I am still finding more. However, to date I have not found any predictions based upon the assumption that mass is basically independent of space and time to be shown wrong by experiment.

Five dimensional geometry

The first implication of the assumption that mass is independent of space and time can be seen in the equation of the first law of thermodynamics above. The right hand side has five independent differentials. This implies we live in a five dimensional universe of space, time and mass. This takes some adjustment in our thinking. More importantly, can we predict something that will be different from experiment?

One of the first things that needed to be done once it was decided to look into a five dimensional universe was to develop the five dimensional equations of motion and the five dimensional fields that produce these forces. This leads back to the work started at the Naval Postgraduate School. This means that one needs to know how the laws of thermodynamics can produce a metric, which houses the information about geometry, and a variational principle to provide the equations of motion.

Fairly early in the study of theoretical thermodynamics the stability conditions are introduced. There is a way of using the first and second laws to investigate how the system may behave if it is displaced slightly from an equilibrium state and then turned loose. If the system tends to go back to the equilibrium state it is termed a stable system. If the system tends to proceed further away from the equilibrium and never returns then that system state is an unstable state. The stability conditions are a collection of second order derivatives. Second order derivatives describe changes in changes of the variables. For example, the velocity is the change of the position with respect to a change in time. A change in this change of position is called the acceleration and this is a second order derivative.

By using the two fundamental laws the stability conditions are the second order derivatives arranged in a certain way and this way provides a description of the geometry in which the motion will occur. This is a very different concept from all prior concepts of geometry in physics. Euclid established the notion and axioms of the geometry that bears his name and in which vectors are straight, of constant length and triangles have 180 degrees as the sum of their interior angles. Newton chose to use Euclidean geometry in which to describe the motion due to the action of his equations of motion. Actually I should not say "he chose" Euclidean geometry. Rather I should say that Euclidean geometry was the only geometry around when Newton did his thing. Newton took a lot of heat for what was called his absolute space and time. Einstein saw that the assumption that the speed of light should be a constant and the same value for all observers traveling at a constant velocity with respect to each other tied space and time together in an unprecedented way. Riemann developed this four dimensional manifold of space-time into a geometrical structure now named after him. In this geometry a vector maintains a constant length but may change the direction it points as it moves around in space and triangles in this manifold do not have interior angles that add up to 180 degrees. Therefore, Riemannian geometry is referred to as a non-Euclidean geometry.

The next step was taken by Einstein when he begin looking for a way of describing the motion of a body under the influence of gravity as motion in a Riemannian manifold that seeks a path of minimum distance between two points rather than motion due to gravitational forces. This means that Newton used the only geometry he knew and developed motion as the result of forces. Einstein sought to find a manifold that was curved by the presence of gravitating mass in such a way that motion follows the minimum distance path without forces. In other words Einstein chose his geometrical type and sought a curved space to fit his choice. It was certainly a good choice and we will see later just how good this choice was. However, we may now see the difference the thermodynamic way is when compared to these two prior examples.

Newton set up his force laws in a geometry, and a space, of his choice even though he only knew of one type of geometry. Einstein used his equivalence principle to establish a curved space in which gravitating mass creates the curvature. Based upon the success of his general theory of relativity the scientific community now claims reality is a curved space-time manifold. Yet Einstein's work does not demand space to be curved. Rather, it describes motion in a curved space without the use of forces. Even in Newton's mechanics the same choice existed. On the one hand the geometry could be chosen and motion determined to be the result of forces. On the other hand there were transformations that allowed the problem to be transferred to a curved space in which the motion was found by the path of minimum distance. This transformation is given in most books on mechanics and so is the transformation that takes the problem back into the manifold where forces determine the motion. However, in the curved space the geometry that appears is neither Euclidean nor Riemannian and, therefore, is set aside as non-geometrical and generally ignored. I have seen no attempt on anyone's part to try to transform Einstein's minimum distance motion in a curved space to a different space where motion is due to forces.

Perhaps Einstein's equations setting up the curved space is simply too elegant for anyone to attempt a transformation that might detract from its elegance. I think we will see later when I show the transformation between a freely chosen manifold with motion due to the force of gravity to the curved space of Einstein's equations, that reverse transformation would have been virtually impossible without considering a five dimensional manifold. Therefore, it would be unlikely anyone would have succeeded even if they had attempted to find such a transformation.

The laws of thermodynamics allow one to freely choose the coordinate system in which they wish to state the problem. Then the laws determine the geometry of the manifold in which the solution may be obtained. Two interesting features of the solution manifold appear. First, the variables of the solution manifold are not unique! Next, there are two manifolds instead of the one manifold expected.

The fact that the variables of the solution set are not unique and may be chosen at will was not too surprising since the stability conditions allow one to investigate the stability using different sets of variables. For example, in classical thermodynamics a description of the system may be made by choosing variables from the set of variables that include pressure, volume, mass density, temperature, entropy, free energy, and enthalpy. Yet only two variables are needed to determine the stability of the purely thermodynamic system with only a single work term. Any set of two variables may be used to investigate the stability. When mechanical systems are considered where there is only one work term, again only two variables are needed but the set of variable now includes force, position (space), velocity, entropy, free energy, and enthalpy. Yes, there are mechanical entropy, free energy and enthalpy functions that are specified by the laws in the same manner as they are defined in the purely thermodynamic system.

I spent a lot of time investigating these various combinations of variables before I found the one that led to a manifold that we commonly use in mechanics. The choice of which set of variables is used may be free, but the difficulty of finding an answer is not equal among the choices. For example, should the velocity be chosen to be in the set of solution variables the very difficult situation results in that the equations of motion are third order differential equations with respect to time. Third order differential equations are difficult or impossible to solve. As might have been expected only one set of solution variables produce the equations of motion that reduce to the old force equals mass times acceleration with which we are familiar from Newtonian mechanics. The fact that only one set of solution variables leads to our familiar equations of motion is not surprising. What is surprising was that the set that lead to Newtonian equations of motion for low velocity systems consisted of space and entropy!

What was even more surprising was that there were two solution manifolds needed for a complete solution. The two solution manifolds are required since there are two things needed to obtain equations of motion; a metric and a variational principle. The metric comes from the stability conditions while the variational principle comes from the second law. Remember that the second law talks about the way heat flows in a purely thermodynamic system and this leads to two principles. One is the principle of maximum entropy principle for isolated systems while the other one is the minimum free energy principle for non-isolated systems. The difference between an isolated system and a non-isolated system is that the non-isolated system may exchange energy, or heat, with its surroundings. The isolated system does not exchange energy with its surroundings and simply increases or decreases its own system energy as it does work or gets worked upon.

Already we see an increase in choices of how the determination of motion may be done that was not available in Newton's or Einstein's approach. One needs to specify whether the system is isolated or non-isolated in order to know which variational principle to use. One also needs to set the problem up in the space-entropy manifold if it is desired to see the usual type of equations of motion. I wanted to find out if thermodynamics would reproduce the classical Newtonian and relativistic mechanics so I did not investigate whether another choice of variables for the solution set might require less effort than the more familiar force equals mass times acceleration approach.

When the stability conditions are written in terms of space and entropy the metric is not recognizable as one leading to the relativistic space-time manifold. Also, the distance between two points in this manifold is not calibrated, or parameterized, in the sense that its arc length is not specified. This may be done by using a velocity multiplied by the differential of time. The only velocity that is well defined at this point is the unique velocity required by the second law. The one velocity that the laws require all to measure no matter what their own speed may be with respect to each other. This velocity is therefore unique and will be the same for everyone. But the variational principle for an isolated system is not given in terms of the time. It is given in terms of terms of the entropy. This means that the metric, in the space-entropy manifold must be solved for the differential change in entropy before the variational principle may be used.

When the differential change in entropy is isolated and squared to allow the use of the variational principle to maximize the entropy a couple of interesting features of the entropy metric appear. First, the entropy metric is a relativistic space-time manifold. Of course this was the objective if there were to be any hope of thermodynamics reproducing the relativistic mechanics. Next, there are two manifolds needed for a complete solution to the motion! There was the entropy manifold and another was an energy manifold. This was not expected and appears as an added, undesirable complexity. This required me to go on a crash course to learn more about geometry because I now needed to learn what type of geometry was required of each of these manifolds.

A little reflection about why the laws require two metrics for a solution to the motion brings to mind that in the purely thermodynamic system one must determine the energy and the entropy for a complete solution. The energy comes from the first law which is a statement of the conservation of energy while the entropy comes from the second law which produces the entropy principle. Thus, in retrospect we should not have been surprised by the requirement of two manifolds of energy (first law) and entropy (second law). Continuing this reflection a little longer brings out the fact that the first law is a non-integrable or path dependent statement while the second law requires the entropy to be integrable (path independent). We should not be surprised then if we were to learn that the energy manifold is path dependent and the entropy manifold is path independent.

While my suspicions about the character of the two manifolds turned out to be correct there was a lot of learning on my part in order to complete the proof that the suspicions were indeed correct. This also required that I learn about an additional type of geometry. Both Euclidian and Riemannian geometries are known to be integrable. The energy manifold should not be either Euclidean or Riemannian while the entropy manifold should turn out to be Riemannian. The new type of geometry, the non-integrable one, turns out to be the Weyl geometry offered by Herman Weyl in 1918 as a means of unifying electromagnetic fields with the gravitational field of Einstein's general relativity. Indeed it was the non-integrability, or path dependence, that led to Einstein's complaint against Weyl's unification proposal. The difference between the Weyl geometry and either the Euclidean or Riemannian geometry is that the length of the vectors are allowed to vary in the Weyl geometry while the length is required to remain fixed in the others. This feature of variable length vectors comes from a function that Weyl named the gauge function. It was the gauge function that Weyl used to show that within his geometry the electromagnetic field is determined by the gauge function. Thus, he argued that the gauge function led to electromagnetic forces and the vector curvature of Einstein's general relativity led to the gravitational forces. Einstein's argument against Weyl's proposal was that the path dependence displayed by the gauge function should show up in a path dependence of the atomic states and experiment showed that atomic states did not depend upon their paths, or histories.

A feature of Weyl geometry that becomes very interesting in this research to find if thermodynamics might require mechanics is that the gauge function provides a scale factor for the manifold. A scale factor determines the length of vectors. It is the gauge function that provides variations in the scale factor that leads to variations in the vectors' lengths. The interesting feature is that should one look at very stable systems, which are systems with constant entropy, the scale factor is required to be a constant value of one, or unity. In 1927 London showed that unity scale factor in a Weyl manifold required motion to be described by Schrödinger's wave equation! Now this was a surprise! An isentropic system was required to satisfy quantum mechanics! Thermodynamics requires quantum mechanics for certain restricted systems! This was the answer to my questions about quantum mechanics.

24

Now it was possible to show that the fundamental laws required that quantum mechanics be used to determine solutions to isentropic (constant entropy) systems. Quantum mechanics was not a fundamental theory, but was a subset of the theory based upon the classical laws of thermodynamics. This caused me to revisit quantum mechanics with renewed interest. I also proceeded to develop the full five dimensional quantum mechanics including the five dimensional Dirac equation.

I would get tired of a constant diet of quantum mechanics so I would vary my research between quantum mechanics, equations of motion and field equations. Both the quantum mechanics and the equations of motion needed forces with which to make the motion go. This is where the field equation research came in. It provided the force laws. You may remember that Newton assumed the law of gravity to give him the gravitational force to use in his equations of motion. Einstein assumed his field equations to give him the curved space within which minimum distance between two points gave the motion. Here, if the laws themselves did not produce forces an additional assumption would be necessary. I did not wish to make additional assumptions. So I followed Weyl's lead where he used the scale factor to derive the Maxwell field equations and from these the force laws. In five dimensions there are four more field components and additional forces. What were these fields and forces?

I could see nothing in the force laws that gave me a clue to the nature of these new forces. When I put these new fields and forces into the five dimensional quantum mechanics there appeared an additional means of generating magnetic moments with the particle spins. Current nuclear physics ascribe anomalous magnetic moments of particles to the effects of nuclear forces so I erroneously thought these new fields and forces belonged to nuclear physics.

There is a theorem in quantum mechanics concerning eigen values that gave an interesting result in five dimensions. Four dimensional quantum spin gives a three component spin vector. In five dimensions this single three component spin vector is expanded into two, three component spin vectors and an additional four component spin vector. I could not determine a good interpretation of the new spin vectors. However, when I applied the theorem to these three spin vectors I found that the three vectors were not independent of each other and that a relationship existed among them such that the components of the three spin vectors only appeared in sets of eight. In two, three component and one four component spin vectors there are a total of ten spin components. However, the theorem showed that the ten components were not all independent, only eight were independent and the other two could be described in terms of the eight independent components. It is interesting that in particle physics particles are found to come in octets. The various attributes of the particles were such that the particles could be put into sets of eight. This is the eight-fold way for which Murray Gel-man received a Nobel Prize. Could this be the origin of the eight-fold way hiding in the five dimensional quantum mechanics?

By the time my two years of teaching at the Naval Academy was up I was deeply interested in continuing this search into the foundations of physics. I asked my detailer for orders to a billet as a military research associate at the Los Alamos Scientific Laboratory. Later the lab's name was changed to the Los Alamos National Laboratory. My detailer was reluctant to give me such a set of orders since it would take me out of a career path that would support advancement in rank. That is he was reluctant until the then superintendent of the Naval Academy, RADM Kinnard R. McKee, gave me a good recommendation for the transfer.

Bunches of Charges at LANL

I arrived at Los Alamos not knowing exactly where I might settle down. The standard procedure for military research associates (MRAs) was that they were interviewed by various group leaders that felt the MRA might fit within their group. I ended up in a group that conducted shock physics where the shocks were generated by high explosives. The primary reason for this choice was that this group had a new data collecting instrument called a VISAR. This instrument used a velocity interferometer to directly measure the velocity of shocks emerging from the surface of a material. In my attempts to find a means of making a prediction that could be compared to experiment I had begun to look at systems of five dimensional flows. This made the capabilities of the VISAR appear to be a means of measuring a prediction not made before.

I earned my keep within the group by applying my knowledge of interior ballistics which I used to help them load their powder driven 20 mm shock gun and in designing and loading unique gun powder driven shock generating devices. At the same time I finalized the derivation of the five dimensional flow equations and showed that the non-relativistic, mass conserved limiting Nervier-Stokes equations contained a non-linear viscous term that did not appear in classical hydrodynamic flow equations. I used this viscous term to explain why the viscosity of shocks in aluminum as measured by Sandia National Laboratory and the Soviet Union showed a dependence upon shock strength. However, I did not get to measure the velocity profile of a shock emerging from a planar material in order to compare this profile with the one predicted with the new flow equations.

An interesting exchange occurred with Dr. Edward Teller during this time. My group leader told me to send a copy of my research on the five dimensions to Dr. Teller's office so that he could see it on his next visit to the lab.

When Dr. Teller came he gave me a call and told me to come to his office so we could talk. When I arrived he quickly came to the point. He wanted me to show him the details of the proof that there existed a unique integrating factor that was a function of velocity only for the first law of thermodynamics when only a single mechanical work term was used. There are three steps to this proof and when I had completed the first two Dr. Teller took over and asked that I let him finish the proof which he did with a flourish. He understood the physics so well that he could immediately see what was needed to finish the proof. He then dismissed me stating that he would call. He never did. He secretary indicated that he did not wish to discuss the research further.

My group leader also told me to send the research to Dr. Sterling Colgate who was there at the lab. I did and soon got a call from Dr. Colgate. He wanted to know why I had sent it to him. When I told him my group leader had suggested it, he indicated that my group leader should have known better. Dr. Colgate then told me, in his usual direct manner, that he didn't understand Einstein's theory so how could he be expected to understand this five dimensional stuff.

The laboratory changed directors and the new director established programmatic oversight offices so as to instill a matrix management system. One of the program offices was aimed at Department of Defense (DoD) programs and the associate director, Dr. C. Paul Robinson, that was to be in charge of this DoD program office thought MRAs should be useful in this office as they should be knowledgeable about how the services worked and what might be useful to them. So I moved to the DoD program office.

Not long after I joined the program office I received a call from an instructor at the Naval Academy who had read the research reports I had written while at the academy and wanted to talk about the research. I later met with him and this meeting resulted in my returning to LANL and getting funding from the associate director to hire Dr. Dan Ross for a year as a research collaborator.

One of the main reasons for hiring Dan was a person who could provide an ability to check the mathematics that I used. This paid a great dividend when he found that in the set of gauge field equations I had reversed the sign of a term. The discovery of this error was like opening the theoretical flood gates. When I made this change in the equations and again asked the question if the quantization that the unity scale factor imposed could be the requirement that specified the quantization of electric charge, I not only found that answer to be in the affirmative but the quantization of the charge added a multiplicative term to the electrostatic potential!

This was not a small change in the theory. It almost changed everything. The classical electrostatic potential was given as the gauge potential time component and was written as

$$\phi_0 = \frac{Z_1 Z_2 e^2}{4 \pi \varepsilon_o r}.$$

The new potential that I derived after I corrected the error Dan found was given by

$$\phi_0 = \frac{Z_1 Z_2 e^2}{4 \pi \varepsilon_o r} e^{\frac{-\lambda}{r}}. \tag{1}$$

The difference between these potentials does not appear to be much until you consider their individual behavior as r tends to zero. Of course, as r gets large compared to λ the two potentials become essentially identical. But when r approaches λ the two potentials begin to diverge rapidly. In fact the classical electrostatic potential tends to infinity as r tends to zero. This feature has caused no end of troubles for understanding nuclear and particle physics and is referred to as a singular potential to denote its tendency to go off to infinity. On the other hand, the new potential reaches a maximum absolute value when r equals λ and then the potential's value returns to zero as r continues toward zero. This type of potential is called a non-singular potential to indicate that it is well behaved for all values of r.

Such a potential is so different from the classical potential that I thought it must surely be wrong for we would have seen its effect in any of several experiments already conducted. So I calculated the scattering cross-section for the non-singular potential and compared the result with the proton-proton scattering results. Since the non-singular potential begins to differ rapidly from the classical potential as r approaches λ the scattering cross-section showed a rapid departure from the predicted proton-proton scattering predictions. The question then arises whether or not such behavior had been seen?

The answer, of course, is yes, absolutely. When the first proton-proton scattering begin to show a dramatic and rapid departure from the expected Coulombic scattering prediction the collective scientific community was put in a precarious position. The Coulombic scattering law was clear as was the data. The experimental scattering was not Coulombic. What might account for this departure from the expected? Could it be that Maxwell's equations were not exactly correct? Surely this was not the case. They were too well established experimentally. Then the deviation had to be due to some as yet unnoticed new force. A force that was independent of Maxwell's electromagnetism. The deviation was due to a force that had a very short range. The scientific community chose the latter and the new force was called the strong nuclear force.

Now reconsider what we have just learned when Dan found my error. Previously I had found that very stable systems had entropy that was constant. Constant entropy requires a unity scale factor. A unity scale factor requires that the electrostatic potential be quantized. This requirement has never been seen to be violated. Now I found that when the quantization is used in the Maxwell equations of electromagnetism it changes the functional form of the electrostatic potential. The change in the potential did not come because Maxwell's equations were changed. The change was due to the quantization of the potential. How can this be? Mathematically there are functions that are referred to as continuous functions. The set of continuous functions that form solutions to Maxwell's equations are described by the classical singular electrostatic potential. When the set of discontinuous functions that are made up of the quantized gauge potentials the functional form of the electrostatic potential receives another multiplicative term that causes the potential to be non-singular and to deviate strongly from the classical singular form for the set of continuous functions. We see from this that the explanation for the deviation in the proton-proton scattering does not come from a change in Maxwell's equations. It does not come from a new force independent of Maxwell's electromagnetism. It comes rather from the addition of the quantized potentials to the set of continuous functions that form the solution to Maxwell's equations.

There is an immediate additional advantage that comes from this new potential. It provides a functional form for the force acting between two protons in close proximity that we never knew about before. We also can use the scattering cross-section prediction and see that for proton-proton scattering the results appear to be near Coulombic until the protons are separated by roughly 10^{-15} meters. This requires that the λ_p for the proton be approximately 10^{-15} meters.

Suppose we now consider electron-electron scattering. Here the data shows that the scattering cross-section still appears to be very nearly Coulombic even as they approach each other to a separation of 10^{-17} meters. This means that the λ_e for the electron must be less than 10^{-17} meters. This means that the point of maximum absolute value for the potential of the proton is much larger than the point of maximum absolute value for the potential of the electron. Now we have a very interesting situation.

A functional form for the strong nuclear force is not the only advantage the non-singular potential gives us. Look at what it has to say about the forces acting between two unlike particles. Say the forces acting between the proton and the electron? In this case the forces do not satisfy Newton's third law which says that for each force there must be an equal and opposite force. The force on an electron due to the presence of a proton is given by the charge of the electron times the electric field of the proton. The electric field of the proton is found by taking the negative of the change of the proton potential with respect to a change in the separation r. Since this means the force is the negative of the slope of the proton potential it shows that the force on the electron goes to zero when it is separated from the proton at a distance of λ_p or approximately 10^{-15} meters. On the other hand, the force on the proton is the charge of the proton times the electric field of the electron. In this case the force on the proton remains strong until the separation between the two particles gets down to the λ_e for the electron or 10^{-17} meters. This sets up the possibility of a proton in a nuclear orbit around an electron. This would constitute a neutron.

This was a surprise, a revelation and, as it has turned out, too big of a shift in physics concepts for the current physics community to handle. Yet every time a neutron decays a proton and an electron always emerge. Currently the physics community states that a neutrino also emerges. The solution to the relativistic equations of motion for the above forces does not require a neutrino to satisfy the spin and energy conservation relations. Not only

32

does the solution satisfy the spin and energy conservation relations, but it also predicts the correct half-life of the neutron. The neutron half-life cannot be predicted using the standard model of nuclear physics. However, the solution of the proton orbit within the neutron requires yet another familiar concept of quantum mechanics to be revisited.

This concept is that of the unit of action. This concept is related to the uncertainty principle which is related to the quantum Poison brackets which in turn are related to the classical Poison brackets. This concept needs to be reconsidered because the classical value of the unit of action, or uncertainty, is such that, as currently interpreted, the uncertainty principle would not allow the electron to be confined to such a small orbit. In mechanics, whether classical or quantum, certain pairs of variables are called canonical variables because they satisfy a certain differential relationship. One pair of canonical variables is the position and the momentum which in quantum mechanics the quantum Poison brackets of the position and momentum equals Planks' constant divided by 2 pi when the geometry in which the differentiation is done is Euclidean This is the unit of action and also represents the maximum of the product of the uncertainties of the position and momentum. This is the Heisenberg's uncertainty principle value. If the geometry is not Euclidean then the geometry helps determine this value. However, Riemannian geometry with only a vector curvature set by gravitating mass such as that found in Einstein's general theory of relativity does not influence the Poison bracket value appreciably. Only the gauge function that scales the length of a vector as in Weyl's geometry influences the unit of action, or uncertainty, on this scale. For gauge potentials such as the electrostatic potential, whether singular or non-singular, the unit of action gets smaller as the orbital size gets smaller. For the orbits of the proton and the electron in the neutron the effective units of action may be found using the magnetic moment of the neutron before it decays and the magnetic moments of the proton and the electron after it decays. The effective units of action in the neutron are determined experimentally to be $\hbar_p = 0.66586\hbar$ and $\hbar_e = 8.0517x10^{-4}\hbar$. With this reduction of the units of action for the constituents of the neutron we see clearly why each neutron that decays produces a proton and an electron.

The particle physicist will be quick to remind me that the standard model says that a neutrino is also to be included in the decay products. The neutrino was first postulated to satisfy conservation of energy because neutron decay produced a change in the motion of the center of mass of the neutron before the decay and the decay products after the decay. Why the center of mass should change its motion during the decay process was unknown. There was nothing acting on the neutron externally to cause the center of mass to change its motion. This prompted the postulation of a neutrino in the decay products to help explain the change in the center of mass motion. Neutrinos were later discovered, first in the presence of a nuclear reactor and later from the sun and later still were thought to be the tell-tale sign of a supernova explosion.

When I first saw that neutron decay could be explained without resorting to having a neutrino in the decay products, I did not know what to do, or say, about the neutrino. Later, I discovered a theoretical prediction of a neutrino that allows the design of a desktop sized neutrino detector. Not only am I now able to design a desktop detector of neutrinos, but I have also designed a desktop neutrino generator. The neutrino is alive and well even if it is not needed to explain neutron decay.

Once I had used the non-singular potential to explain the neutron I was eager to see if the non-singular potential allowed me to develop descriptions of other more complicated nuclei. The exponential character of the potential made it so difficult that I did not obtain solutions at that time, but I did derive the energy equation for a deuterium nucleus by using the symmetry of two protons in orbit around a single electron. Next I derived the equations for three protons in orbit around the electron and also for four protons around an electron. I could not tell if these represented helium 3 and lithium 4, but I could see that if somehow there were to be a nuclear core with two electrons the four proton equation would represent the helium nucleus. Since I didn't know how to describe the negatively charged core, I used the masses of different isotopes, while assuming that the core energy was identical among isotopes, to determine a value of energy for the core of each nucleus. I then applied the same procedure used in atomic physics of putting additional protons in additional energy shells formed by higher quantum numbers until I was able to predict the masses of 21 nuclei, including isotopes and excited states, up to a Z of 4 and a mass number of 10. In constructing this table of predicted masses I only used the energy equation for the two proton symmetric case and just added additional protons in higher orbits. This necessarily meant that the errors in predicting helium 3 and helium 4 would be greater than if I had used the three proton and four proton equations. Even with these unrealistically large errors the RMS error for all 21 nuclei masses was 2.9 MeV. This accuracy may not seem too good until compared with semi-empirical mass formulas that do not address nuclei below an A of 16.

Dr. Paul Robinson asked me to present this nuclear model to the theoretical physics group. He had previously asked me to present the non-singular potential to them and their response was a guess that such a potential was far too simple to explain the complex structure found within even the low Z nuclei. Robinson now wanted to see what the theorists would say about the first model that was constructed from the potential. Their response was that this was far too complicated. They liked their way better.

When I tired of nuclear physics I applied the non-singular potential to other realms of physics. I wondered if the non-singular potential, by the very fact that it was different from the Newtonian gravitational potential, might explain the advance of the perihelion of the planetary orbits. I used the technique of looking for small radial oscillations about steady circular motion and found that if the λ_S for the sun had a value of

$$\lambda_S = \frac{GM}{c^2}$$

the predicted perihelion advances were identical to the predictions of Einstein's general theory.

I also used the five dimensional gauge field tensor to determine the inductive coupling between the electromagnetic and the gravitational fields. I did this using the fact that if both the electromagnetic field components and the gravitational field components were all included in the same tensor equation they would all need the same units. I found the charge-to-mass ratio that allowed all components to share units was

$$\beta = \sqrt{\varepsilon G} \, .$$

Here ε is the dielectric constant and G is the gravitational constant. For free space this coupling constant is 2.4296×10^{-11} coul/kg.

The coupling constant between the electromagnetic and gravitational fields allows one to predict phenomena not previously possible. I first remembered that in the five dimensional quantum mechanics that a purely gravitational mass (that is electrically neutral) had a magnetic moment. I wondered if I could predict the magnetic moment of the spinning body called the Earth. So I applied this ratio to the Earth and, if one assumes the mass of the Earth is uniform

36

in density, the predicted magnetic moment is 8.6×10^{22} amp-m^2. This prediction compares fairly well to the experimental value of 8.1×10^{22} amp-m^2.

Another field of inquiry was given to me by Dan Ross. Dan had been intrigued by the large red shifts from quasars. These stellar objects had earned their name by having such a large red shift that put them among the stellar objects that were far removed from the Earth and yet they were very bright for their distance. The unanswered question was how could they be so far and so bright? So I looked at the red shifts of the non-singular potential.

The calculation of the predicted red shifts also included a feature of the gravitational potential not shared by the electrostatic potential. In the five dimensional gauge field tensor the structure of the tensor requires that the gravitational potential and field be a function of time. I couldn't find the set of equations that would tell me what this function was. So I assumed a linear form for the time dependence. Also, you may recall that I showed that the unit of action depended upon the gauge function. Einstein had shown that the photon energy was given by $\varepsilon = h\nu$ so the energy of the emitted photon also depended upon the gauge function where it was emitted and the frequency of the received photon depended upon the gauge function were it was received. This meant that the red shift depended upon three things; the distance between where it was emitted and received, the gravitational field strength were it was both emitted and received and the time of travel of the light from the time it was emitted till the time it was received.

Hubble saw a dependence of the red shift upon the distance. Einstein's theory predicts that in addition to the distance the red shift also depended upon gravity. Now here is a prediction that also includes a dependence of the red shift upon the time of the light travel. Now you might say that the distance between the star and the Earth is also a measure of the time and you would be correct. But that measure of time says nothing about how much stronger the gravitational field of the stellar object was when the photon was emitted than it is when the photon is received. Yet the biggest factor in the red shift that applies to an understanding of quasars is the exponential part of the potential. We saw that the exponential part depends upon the mass of the object when we evaluated the lambda of the Sun by comparing the prediction to the experimental planetary orbits. If we call these three parts of the red shift the distance, gravity and time parts, we see that the gravity part has the largest exponential effect and the red shifts of very massive stellar objects will have large red shifts due to the exponential part way over and above what Einstein's theory would predict.

In addition to this explanation of the large red shifts for quasars the fact that the red shifts depended upon the gravity where the light is received these expressions showed that quasar red shift received at the Earth's surface would be different than the red shift from that same object that is received at the altitude of the Hubble Telescope. This prediction excited Dan and he went off to talk to folks at the Hubble Telescope. He was extremely disappointed when he found that they systematically calibrated their data to the Earth so that any difference was screened out! If the data were not recalibrated to compare with the readings at the Earth's surface the difference between the readings would allow a prediction of both the distance to the object and its mass. This interesting information is lost due to the recalibration process.

Another realm of physics that Dan and I spent a lot of time researching is waves. Not the waves that you make on ponds. Since the theory predicted an inductive coupling between electromagnetic and gravitational fields, then there should be electromagnetogravitic (EMG) waves. This is a big word, but I use it to try to indicate the inductive coupling that exists between the fields. We know that static electric and magnetic fields may appear without the other field being generated. We also have worked for years with gravitational fields without suspecting that they are coupled to the electric and magnetic fields. Well now, this is not entirely true. This phenomenon has been suspected by a few researchers in the past starting with the great electromagnetic researcher Faraday, but their efforts have not been generally accepted by the scientific community.

The first step in looking into EMG waves was to derive the wave equations. When I did this I obtained four wave equations. Just as in four dimensions, I had a wave equation for the electric field and a wave equation for the magnetic field that were coupled together. Each of these equations contained a term that involved how the field changed in space. They each had two terms involving the rate of change of the field in time; one was linear and the second quadratic. These terms were identical to the four dimensional Maxwell wave equations. Each of these wave equations had one additional term that involved the change of the field with respect to mass density. The newness of the concept of how, or even why, the electric or magnetic field should vary with changes in the mass density made visualizing these terms almost impossible. Yet there was no denying that if the fields had a five dimensional basis these terms should be added to the wave equations.

But I found two additional wave equations were necessary. One of these was another vector wave equation similar to the electric and magnetic wave equations complete with the four types of terms each of these equations contained. The gravitational field wave equation had two additional terms that had no counterpart in the electric and magnetic equations. These two might really be considered a single term as they both involve the change of the electric field with respect the mass density linearly. This is a curious way for the electric field to influence the gravitational wave, but it is similar to the manner in which the linear time rate of change term appears in each of the wave equations. The coefficient on each term involves the conductivity of the medium the wave is passing through.

The final wave equation is the odd man out. It is not a wave equation for a vector. It is a scalar wave equation. The literature contains a lot of speculation about scalar waves. Mostly the speculation appears in science fiction and dire predictions about weapons. Here is the first time I have seen a derivation of a scalar wave equation. It looks just like the vector wave equation for the electric or magnetic fields but has a scalar quantity as the subject of the wave rather than a three dimensional vector.

This is the field wherein I really appreciated the help of Dr. Dan Ross. His degrees were in electrical engineering and most of his industrial employment had him working with electromagnetic waves. Dan was one individual that I have known that could most readily set aside his biases while he considered the implications of a new proposition. Most of us cannot do that and keep the implications of our bias from coloring our view of the new proposition. Dan was very good at keeping bias in check; except in waves. Here Dan struggled with his long years of experience with electromagnetic waves. At the same time I speak of the trouble Dan had with his biases, I must also add that this was the realm of research where I appreciated Dan's help the most and I shared his bias and trouble with it. My bias towards electromagnetic waves will show up later.

I struggled with my attempts to find solutions to the set of wave equations. The equations were difficult and the concepts were new. This combination wrecked havoc on our confidence in our derivations and attempts to find solutions. However, early on I was able to show that the solutions appeared as two sets of solutions. Basically the split in the sets of solutions was a split in the types of waves predicted. One set of waves were the type of wave that have been called transverse waves since electromagnetic waves have been studied. This term is used to indicate that the direction the vector in the wave points is perpendicular to the direction the wave is traveling. This is like the waves you see spreading out on the pond when a rock is thrown into the water. The wave travels in a horizontal direction while the motion of the water that you see is motion in the vertical direction. In the electromagnetic waves both the electric and magnetic field components of the wave are directed perpendicular (90 degrees) to the direction the wave propagates. The transverse set of the EMG waves have three vector field components; an electric, a magnetic and a gravitational component.

One of the first things that I thought of when I saw that additional component in the wave along with the electric and magnetic component was that the energy contained in the gravitational part of the wave must be small or it should have been noticed in one of the many experiments that have been done on electromagnetic waves through the years. So I looked at numerous types of experiments seeking a prediction that might be seen in some experiment. When I derived the equations that predicted the energy density and the radiation pressure of the wave, I found a possible experimental detection of the gravitational component. The energy density of the wave is the sum of the squares of each wave component while the radiation pressure is the sum of the squares of the electric and magnetic components minus the square of the gravitational component. Now I had a predictable difference. Had such an experiment ever been done that compared the energy density against the radiation pressure of the wave? The answer was of course, Nicholls and Hull had done such an experiment a long time ago. Their data showed that the radiation pressure was less than the energy density, but within their experimental error.

I then sought a means of conducting a new radiation pressure/energy density experiment with lower experimental error. I talked and planned experiments with several individuals in the lab that used high energy lasers. The problem was in the errors of measuring the energy density of lasers. Because the method of measuring the energy density of light basically includes absorbing the light and measuring the increase in temperature of the absorber, even the best equipment available had an experimental error as great as that of the Nicholls and Hull experiment. So we still could not get better determination of the ratio of the two quantities. One professor friend of mine from the naval academy designed an experiment that used an interferometer to measure the temperature of the absorber. The predicted experimental error of such a means of measuring the energy density was low enough to see the predicted difference, but several attempts to get the funding to conduct the experiment failed to find any source of funds. This is an experiment that should be done.

Perhaps there was another way of seeing the predicted difference. The expression for the energy density was the sum of the squares of all components while the radiation pressure was the sum of the squares of all but the gravitational component which had its square subtracted from the rest. This meant that one could find a frequency of light where the radiation pressure was zero, but the energy density was still positive. Could such conditions be found or created? Such conditions exist if the space of the universe is considered. For example, if one considers outer space and asks how a positive pressure might be maintained it is difficult to find a physical means by which a positive pressure can be sustained. We know that if we have a container here on Earth that the walls of the container can sustain a positive pressure within it. But I cannot think of what physical phenomena might constitute the walls of outer space that could maintain a positive pressure in space. Therefore, I asked the question of whether or not a positive energy density could be measured assuming the radiation pressure of space was identically zero? The answer is that it has already been measured. It may be seen in the cosmological background radiation. Here is a small positive energy density with no apparent radiation pressure. My professor friend from the naval academy did a quick estimate from the Nicholls and Hull experiment, assuming the difference was real, and argued that those results compared favorably with the 2.7 degrees of the cosmological background radiation. But this flies in the face of the beliefs all that those who hold to the big bang model of universe expansion.

Let's go back now to the second set of waves that forms solutions to the wave equations. In order to compare these with the transverse waves lets refresh our memory of the transverse wave by recalling that each vector component of the transverse wave is perpendicular to the direction of the wave propagation. The magnetic component is perpendicular to both the direction of wave propagation and the electric component. The gravitational component is perpendicular to the direction of wave propagation and opposite in direction to the electric component. This means that if you are looking in the direction of wave propagation the electric component is directed up, the magnetic component is to your right and the gravitational component is down.

Compare this to the other set of waves that I call the non-transverse because none of the components of this wave are directed perpendicularly to the direction of wave propagation. However, not all of these components are in the direction of the wave propagation so I didn't feel I could call them longitudinal waves. One component of this set of solutions is a scalar component. It has no direction and, therefore, could not be said to lie in the direction of wave propagation. For this reason I prefer to call this set of solutions the non-transverse waves. They consist of an electric component directed in the direction of wave propagation, a gravitational component opposing the direction of wave propagation, and a scalar gravitational component that has no directivity, but must accompany the electric and gravitational field component.

There is an interesting feature of the non-transverse waves. Transverse waves have an electric field component directed perpendicular to the direction of wave propagation. Therefore, such a wave, when it impinges perpendicularly upon a conductor so that the electric field component is directed along the conductor and can force electrons in the conductor to move. This is why radio waves can be picked up by an antenna. Also, an electron that moves back and forth in a conductor can generate an electromagnetic wave in the reverse of the receiving antenna. This is the heart of the radio, cell phone, and television industry. Because the non-transverse waves would not have an electric component in the direction of the conductor, I see no way for non-transverse waves to be picked up by the transverse wave antennas. In other words transverse antennas could not detect the presence of non-transverse waves.

This is the reason that Lockheed Skunk Works gave me a contract when I retired form the Navy. As Ben Rich, director of the Skunk Works, told me when I was called out to begin this contract, a stealth fighter might not be seen by radar, but as soon as the pilot keyed his radio everyone in the area knew he was there. Ben Rich wanted a stealthy communication system for the stealth fighter. Though I worked hard on trying to develop a set of antennas that would transmit and receive the non-transverse waves I failed. It was only later, after Ben Rich died, that I learned that my failure was due to my bias with regard to antenna design that prevented my success. Remember I mentioned

Dan Ross struggled with his bias. Here I didn't even realize that my bias prevented my success for several years. But we will return to how to design the stealthy communication system later.

By now my research had touched every branch of physics and still I had not found a prediction that differed from experiment. I had found several that differed in interpretation from the standard model. But a difference in interpretation should not constitute, and is not, a disagreement with experiment. Indeed several predictions more closely aligned with my notion of common sense than did the accepted interpretations.

Big Bangs at New Mexico Tech

The State of New Mexico established five centers of technical excellence within the state to encourage additional research and industrial development within the state. One of these centers was the Center for Explosives Technology Research (CETR) at the New Mexico Institute for Mining and Technology, informally called New Mexico Tech, in Socorro, NM. The university had hired a director, Dr. Per-Anders Persson, from Sweden who had previously had a distinguished career as director of Nitro Nobel's research center. Dr. Persson had heard of my work at Los Alamos National Laboratory with respect to explosives and shock physics and contacted me to see if I could be persuaded to go to Socorro to help him establish, develop and administer CETR. This started the next phase of my research.

As may be obvious by now the theoretical research discussed herein was not my main occupation. Only twice did it make up even a portion of what I was paid to do. Once while I was at the naval academy where the faculty thought it was so unique to see a naval officer interested enough in physics research that the usual requirement for naval officers assigned to the naval academy to spend their summers with the midshipmen was waived and I spent my summer conducting research. Again, while at LANL Dr. Robinson allowed me to conduct whatever research I desired with half my time while the other half I supported the lab DoD program development and administration. Though this may seem to be very generous on his part, it must be pointed out that the lab did not pay my salary. The Navy still paid me so whatever I did for the lab was a freebie in terms of my salary. They provided my support with an office and such.

When I first got to CETR I was pretty well burned out with theoretical physics in that, though I had shown one could explain phenomena a different way and even predict new phenomena, the physics community held their beliefs very dearly. Not only that but they seemed to resent outsiders or those they did not feel had paid their dues. This can be seen in the advice one of the senior fellows at LANL gave me. I had sent some work to him and he looked it over and called for me to visit him. When I got there he quickly told me in his normal straight forward way that, "If you want me to shoot you down; I can't. If you want me to help you; I won't. My advice is that you learn to play the game and then maybe someone will listen." It had never entered my mind that theoretical physics research was a game that was to be played by the rules set by the accepted members of the community and I told him so. In retrospect I should have known that the physics community is made up of a lot of humans and should, therefore, be expected to act like humans rather than totally objective scientists with no human traits. Shame on me.

Sometime after I got to New Mexico Tech the physics journals began printing articles about a fifth force of nature. It seems there were some tests of the Earth's gravitational field done in deep mines and in submarines underwater that indicated that gravity was not quite what Newton's gravitational law said it should be so a new force

46

was postulated to account for the disparity. I felt it was time to compare their experimental findings with the non-singular potential.

One difficulty with the non-singular potential that had made me reluctant to use it for predictions was that the exponential character of the potential made it difficult to find solutions. In addition, when I begin to apply it to the problem of determining the gravitational affect to be predicted in a deep mine this exponential character stuck up an additional ugly head. The deep mine problem can be approached by considering the gravitational attraction of two objects. First one may consider the gravitational attraction of the sphere under the point of the experiment down in the mine to have all the mass concentrated in the center of the sphere. However, when one integrates over all the mass of the sphere to learn its influences on a test mass at the point of the experiment one learns that the exponential term in the non-singular potential requires the gravitational attraction to be slightly different than Newtonian gravity. Also, when integrating around the mass that forms a shell around the Earth with a thickness of the depth of the experiment the result is also a little different from Newtonian due to the non-singular feature of the potential.

I had one of my students do a survey of all the sub-surface data and these compared favorably with the non-singular potential prediction. However, by this time scientists had reported the results of similar experiments up above the Earth's surface with no deviation from Newtonian gravity. This effectively killed all further publication of fifth force articles since it was argued that any deviation from Newtonian gravity should be seen the same above as below the Earth's surface. That argument is valid when the classical singular potential is used. When the non-singular potential is used the above ground data should show Newtonian gravity while the sub-surface experiments will show a slight deviation. Thus, it seems that both the above and the below surface data were compatible with the non-singular potential, but were not compatible with the singular potential. By this time, though journals would no longer consider articles on the fifth force even though the article was arguing that all the data was good just the interpretations were wrong and no fifth force was needed.

When the first reports of cold fusion hit the news a friend called me and asked if I had applied the non-singular potential to the cold fusion case. I told him I had not and did not intend to do so. He thought I was missing an opportunity and indicated that he had applied for a patent on three aspects of the phenomena that the original scientists had not seen. This did well for him for later his patent claims were purchased and he quit his job teaching at a community college and bought property in the mountains of New Mexico where he established his own laboratory and has been having fun ever since.

One of our clients at the Energetic Materials Research and Testing Center, (EMRTC) a center formed by merging CETR with a preexisting Terminal Effects, Research and Analysis (TERA), asked me to look into the phenomenon. I began to consider how the nuclear model I had developed at LANL might be used to describe nuclear fusion. I began looking at how to calculate the fusion potential of two deuterons with the knowledge that a deuterium nucleus consisted of two protons in orbit around an electron. This was a six-body problem that involved transcendental equations (equations with exponential terms). Though I found this next to impossible to do I got my friend to make me some test material and I used a small amount of funds that New Mexico Tech returned to me on programs that I had brought to the university to conduct some experiments. About the best I can say of the data from these experiments is that they were tantalizingly

48

positive. Now I had to do a better job of solving the six-body problem.

I knew the forces within the deuterium nucleus did not satisfy Newton's third law and the relativistic quantum mechanical equations that were needed in the solution depended upon the assumption that the forces did obey Newton's third law. The only way to approach the problem, therefore, was to redevelop the equations of quantum mechanics for forces that did not obey Newton's third law.

To help show why this was needed it may be useful to indicate just how useful Newton's third law has been for the scientific community all these years. Newton's third law, as you may recall, is the one that states that for each force there exists another force that is equal and opposite. This makes solving a two-body problem much easier because this law can help by reducing the two-body problem to a one-body problem. If the forces do not obey Newton's third law we cannot reduce the two-body problem to a one-body problem. I had no choice but to redo the whole of mechanics showing how one solves problems when Newton's third law does not hold.

Redoing the whole of Newtonian mechanics and both non-relativistic and relativistic quantum mechanics turned out to be a laborious, but useful, exercise in fundamental mechanics. The most interesting aspect of this exercise was the best way to set up the problem. For example, if there are two bodies, number 1 and 2, then we may set up our equations to determine the motion of each particle. When we do this showing that the resulting equations reduce to the classical equations for forces that obey Newton's third law becomes difficult. On the other hand, if we set up our equations to determine the motion of the center of mass and the motion about the center of mass the reduction to classical equations for Newtonian forces becomes obvious. So that is the way I went through the redevelopment of mechanics.

First I redid Newtonian mechanics and then moved on to quantum mechanics where I first redeveloped the non-relativistic equation before I redid the relativistic equations. As I indicated, this process was somewhat tedious. It was an excellent exercise in maintaining the real fundamental laws in mind and why we did, or do, certain things in mechanics. The result was very enlightening. I was able to put the relativistic circular orbits of the deuterium nucleus into a spreadsheet and numerically solve the three-body problem which is the stable deuterium nucleus. I was not able to solve the six-body problem at this time.

When I read the news that Ben Rich had passed away I was grieved once more that I had not been able to develop the communication system he had wanted for his stealth fighter. This problem began to occupy more and more of my thinking until one day I was struck with a new concept. Perhaps the concept was not new, but the words with which I began to think of antennas were new. The essence of the concept starts with the realization that, as we noted above, a moving electron creates a transverse electromagnetic wave. On the receiving side, the electric field in the transverse wave impinging on an antenna causes an electron to move inside the antenna. The transmitting antenna is a particle moving to establish a field and might be thought of as a particle-field antenna. The receiving antenna has a field moving a particle which might be called a field–particle antenna. In both cases a particle with inertia is involved. Particles with inertia cannot respond to fields that are directed in the direction of wave propagation the way particles with inertia can respond to fields that are directed perpendicular to the direction of wave propagation. We need to develop a field-field antenna because waves can respond more rapidly to fields than particles can.

I had been thinking of antenna design as I had been taught and this would not work with the non-transverse waves. I needed to think differently. I needed to come up with a field-field antenna.

When light strikes a discontinuity in the medium in which it is propagating a couple of things happen. Some light is reflected at a given angle. Some light is transmitted at a different angle. While the details of this phenomena may not be understood, the phenomena lends itself to being understood and predicted using bulk material properties such as the index of refraction and material density. Of course, we know the index of refraction is the results of the detailed interaction of the material properties such a dielectric constant and the magnetic permeability. The point is that at such a boundary a whole set of boundary conditions must be met. These conditions include such requirements as each field component must be continuous across the boundary.

The question that came to my mind was could I use boundary conditions at an interface between two media, such air and glass, to change some of the energy of a purely transverse wave into a non-transverse wave? To answer this question I had to derive the reflection and transmission coefficients for all ten reflected wave components and the ten transmitted wave components. In doing this I ran into several totally new concepts. For instance, there was a gravitational magnetization to add to the Amperian magnetization. Also, there was a gravitational susceptibility in addition to the magnetic susceptibility. Additionally there was a concept of gravitational induction and a mathematical property that I don't know how to name. But when I was done I was able to show that by choosing the material properties, angles of incidence and polarization some of the energy in a transverse wave could be converted into energy in a non-transverse wave and back again. Now

I had the basics of a communication system that could not be detected by any of the transverse antennas.

You might think the military would be interested in such a communication system and they were and are interested. They are just so dependent upon science advisors who are trying hard to keep the military funding sources out of trouble by steering them away from frivolous research that I could not find funding anywhere. Oh, I long for the days before the Proxmire Golden Fleece awards.

I still had not really gotten anything published and in thinking about what I needed to do to get something published I begin to wonder what the publishers would do if I could show that Einstein's general theory of relativity was but a subset of a more general theory. Also, it might help if I could derive an explanation of the wave-particle duality which had caused Einstein so much concern for the last two thirds of his life. What if I could derive the Yang Mills equations for the weak interactions? Or, what would happen if I could derive the equations for the strong interactions? If I could show that these all were subsets of a more general theory could I get it published? I probably would need to start from some point other than from the laws of classical thermodynamics so that the thermo stigma would not be applied. The thermodynamics required Weyl's Gauge Principle, which Weyl used to derive Maxwell's electromagnetism. I could just start from Weyl's Gauge Principle and not bring thermodynamics into the picture until after I had completed the desired derivations. I could also just assume the existence of a fifth dimension that was physically real without stating what it might be until later, if I could then derive Einstein's general theory.

So when I got to the office one morning I cleared off a spot on my desk and put some fresh paper there. Then I said a simple prayer asking God to help if He were so inclined. In fifteen minutes I had derived a wave-particle requirement for light. Years earlier I had asked what electrostatic potentials where possible for isentropic systems. This time I looked at a different potential. In the five dimensional manifold the electrostatic potential is the gauge potential for the time dimension. This time I chose to look at the gauge potential associated with space. This potential is called the vector potential and is related to electromagnetic waves. Transverse waves can be oriented so that the fields are pointing a certain way. This is called polarization and I used it here to help simplify the equations. The Weyl Gauge Principle requires that the vector potential be a wave, but imposing the isentropic condition also requires that the wave be quantized. This means that there exists waves that are not quantized, but a lot of waves must be quantized. It also means that those waves that are quantized travel in a stable fashion with constant entropy. This is a property of photons. However, there was a new twist. The photon energies were given by the expression

$$\varepsilon = N^2 h\nu \,. \tag{2}$$

Einstein used a statistical approach and gave an energy for the quantum of light to be the expression in Equation (2) with *N=1*.

I wondered if this feature of photon energy might not have been seen in laser light so I asked some of the laser experts if anyone had ever looked at the frequency of the laser output to see if any energy was being generated with frequencies equal to $1/N^2$ times the fundamental frequency? Their first reaction was why would anyone do that? Their second comment was to admit that since no one expected any output in frequencies that far removed from the fundamental plus the fact that detectors usually measure a much narrower range of frequencies that they doubted any one had ever bothered to look that far from the fundamental frequency.

Next, I was advised to check the literature looking at the original data on the frequencies that were measured for excited atoms. When I found the original article at the National Technical Information Service I found that indeed there had been reported outputs for the N=2 and 3 for both hydrogen and helium and these were reported to be in the right proportion relative to the fundamental frequency! This meant nature has a whole new untapped bunch of lasers. I called the new photons phat photons because their energies were larger for a given frequency than Einstein photons. I went on to derive the laser relations for phat photons and their scattering cross sections to see how well they might propagate through aerosols in the atmosphere above the ocean surface, where the navy is interested in using laser weapons. I found the scattering cross sections were reduced by the factor of $1/N^8$. This is a significant reduction in the loss of energy of laser light passing through the atmosphere! I looked at the output relations of the free electron laser to see if this type of laser could be made to emit a chosen phat frequency. I found that there were parameters that could be used to control the quantum number N of the output photons. Though Northrop Grumman continues to attempt to get the US Navy to start up a free electron laser program the war in Iraq is preventing a program startup.

After this productive session, I again went to the office and asked God for help in deriving Einstein's general theory of relativity from Weyl's Gauge Principle. Again within less than a half hour I had a completed derivation. What I did was to assume that there exists a fifth dimension that was physically real, but that I did not know what this dimension was except I required that this fifth dimension be conserved. This conservation requirement puts a restriction on the five dimensional manifold that creates a four dimensional hyper-surface within the five dimensional manifold somewhat like a surface that is a constant radius from a given point embeds a sphere in a three dimensional manifold of space. I found that this restriction required a description of the curvature of the hyper-surface to be given by a set of equations that turned out to be Einstein's field equations if the fifth dimension was the mass density. This meant that Einstein's general theory is the four dimensional hyper-surface that conservation of mass embeds within a five dimensional manifold of space, time and mass. This is just what I had derived from thermodynamics, but without using thermodynamics.

Again I went to the office early and again asked God for help in deriving the Yang Mills equations for the weak nuclear force. Again the answer was quick in coming. This time it required recognizing a certain pattern in the equations describing the motion of particles in a two-particle system using the relativistic quantum mechanics under the influence of gauge fields. Once this pattern is recognized a definition can be made that casts the equations into the Yang Mills form. I haven't done anything with these equations as I find it easier to use the non-singular potential in the other format.

It was a small step to then look at the three-particle systems and recognize the pattern that allowed me to put the three-body equations into the form currently used for the strong nuclear forces. As with the Yang Mills equations I do not see an advantage in using these equations. These equations were derived because the standard model does not have a functional description of a potential from which to get a functional form for the force and they were trying to find something to help learn about the nuclear forces. Once a potential has been found the force laws themselves are the best way of finding the motions.

At this point I wrote up what I had obtained. I put these last derivations in the first part of my article and then I put the thermodynamic derivation of Weyl's Gauge and Quantum Principles in the next section. I felt that in this fashion the reader could see that there was one common point, Weyl's Gauge Principle, from which all currently accepted mechanics could be derived. Along with these derivations I showed that new physics could be obtained. This article was published.

Retirement does not mean Retiring

Once I retired from New Mexico Tech I began a lot of consulting work. This took a lot of my time, yet it left me with time to pursue physics a little more than in the past. I tried to push Northrop Grumman to get funds to work on the alternate communications or the free electron laser projects. I presented papers at different conferences to get exposure for my work. However, these efforts have not had any real payoffs. Northrop Grumman did hire a scientist to investigate my prediction of inductive coupling between the electromagnetic and gravitational fields. His experimental results to date have supported my prediction solidly.

I continued to try to solve the six-body problem that would allow me to calculate the fusion potential for a deuterium-deuterium to helium fusion reactor. The best I have done to date is to solve the problem in a spread sheet. This is a rather crude solution, but it does allow me to calculate the fusion potential to the point where, if the calculations are close to reality, a reactor may be designed of various sizes for different applications. I presented these results at a conference in Albuquerque. Not long after that presentation I was approached by the navy to guide them in an experimental effort to test the fusion concept. This effort is on-going.

My good friend James O. (Oke) Shannon suffered his second heart attack and seemed to be depressed so I attempted to get him interested in technical stuff. I thought I had succeeded when he called me one day to ask me if I had applied the theory to the dark matter and dark energy issues. He felt strongly that these hypothesized dark things were unreal. I thought by helping him look into the dark side of cosmology I would help him along.

When I began to look into the dark matter data, I found that the scientists were looking at the tangential velocities of spiral galaxies and comparing these velocities to the velocities that would be predicted by Newton's gravitation. They didn't match. Newton's gravity argues that as one looks farther out along the arms of the galaxy the tangential velocity should increase until almost all of the mass in the galaxy is contained at smaller radii. Then the tangential velocity should fall off as you look farther and farther outward. The data showed very clearly that this was not the case. The tangential velocities remained approximately constant. The only way the community could make sense of these data was to suppose that there existed additional mass in outer regions of the galaxy that we couldn't see. This is the dark matter that is required to account for the tangential velocity data. The measured velocities argued that a lot of dark matter is needed.

I found that if I used my time dependent gravitational potential and field the data was explained. It runs like this. Spiral galaxies are big things. It takes a lot of time for light to travel from the center of the galaxy out to the arms where the tangential velocity is being measured. In this length of time gravity weakens a lot. However, the stars in the arms of the galaxy do not respond to what gravity is when light can reach them after starting at the center of the galaxy. They respond to what gravity was at the time the light left the center of the galaxy. The speed of light is the fastest that any change in the gravitational field can travel so the stars in the arms are only responding to the gravitational changes as they see them. Since the stars are responding to a gravitational field strength that is greater than current researchers calculate with a time-independent gravitational field the data does not match their predictions.

The upshot of this is that the data of the tangential velocities in the arms of spiral galaxies are what a time dependent gravitational field would have them be. No dark matter is needed.

Next I looked at the dark energy issue. This is a slightly different, but still a similar issue. Scientists have been trying to find two different methods of obtaining distances to distant stars. One way of obtaining a distance is to look at the red shift of the light from the star. By using Hubble's linear relationship the distance would be proportional to the amount of the red shift. Another way of obtaining a distance is to look at the amount of light we receive from the star and compare it to the amount of light to be expected at a given distance. This is like taking a standard 100 watt light and starting down the road. Your friend can stay behind and calculate how far you have gone by the intensity of the light he is receiving. In order to translate this method to the stars the scientists need to find a standard light, or candle as they call it. Some time ago Chandrasekhar calculated that a star that is increasing its mass by drawing in material from around it keeps getting more massive and yet is being held against collapse by pressure created by the particles within the star itself. As the star increases its mass, the strength of its gravity increases until the mass reaches a point at which the star collapses. This creates a bright explosion that is a super nova. The mass at which this collapse occurs is called the Chandrasekhar limit and this limit is thought to be the mass of all exploding type Ia super nova. This could provide the standard candle the scientists need since all exploding stars with equal mass would put out the same amount of light energy. Now all they need to do is to identify the type Ia super nova and measure the light received and compare it to the light output at the star to get a measure of the distance to the star.

Here too the scientists have assumed a time independent gravitational field. The time dependent gravity takes away their assumption of a standard candle in this fashion. A long time ago gravity was stronger. Less mass was needed to create a certain gravitational pull. It would have taken less mass to overcome the electron pressure within the stars. This means there would have been less mass to generate light upon explosion. This means that young super novas have more mass and, therefore, more light output than the older super novas. This would mean that, assuming a standard candle, one would think that the older super nova were farther away than they should be. This is precisely what the scientists have concluded. They think the data is telling them that older super nova are dimmer than they should be for a universe whose expansion is slowing down and, therefore, the universe expansion must instead be speeding up. They conclude that this must be due to some mysterious energy that we can't see and this is the dark energy reference.

On the other hand, if indeed gravity is getting weaker as time goes along then the older super nova had less mass than the newer super nova and the data is correct even with the universal expansion slowing down. Thus, the data currently used to support the dark energy hypothesis is also support for the time dependence of the gravitational field.

Now I want to go back to the question of neutrinos. You may remember that I was able to describe the decay of a neutron without the need for a neutrino. Yet neutrinos are thought to have been, and are being, measured. You may also remember that I first looked at the electrostatic potential and found that the Weyl Quantum Principle quantized the electric charge and the associated electrostatic potential. Later I looked at the Weyl Quantum Principle influence on the gauge vector potential and found that it quantized the energy of a photon and even showed that the energy of the photon had a quantum number. There is one gauge potential left in a five dimensional universe and that is the gauge potential associated with the fifth dimension and gravity. I looked at what happened when the Weyl Quantum Principle was applied to this potential. The result appears to have all the attributes of a neutrino. It has an energy that is proportional to the square of a quantum number and frequency just as the phat photon did. It does not have a polarization capability as the phat photon did and that is appropriate for a non-transverse wave for only the transverse wave cares about the direction the fields are pointing to the side.

This is a rather interesting prediction of a neutrino. If this prediction is true then the alternate communication system is a system that communicates by neutrinos. Also, if this prediction is true we can make a flux of neutrinos and aim them at the current neutrino detectors and see if the alternate transmitting antenna really does make neutrinos. How does nature create her neutrinos? I don't know yet, but it will be interesting to find out how she does it.

Where does this leave us? What have I really done?

My thesis advisor told me early on that if I was right then they were all wrong and the whole book of physics would have to be rewritten. I think it would not be fair to them, or me, to say that I have rewritten the whole book of physics. I think I have made a significant alteration or modification to it. Do I think what I have done is correct? With so many positive comparisons to experiment I find it hard to think it can be wrong. Still I wrote this because I am now waiting to witness experiments to be conducted in a few days the Naval Surface Weapons Center to test the fusion prediction and I have been getting anxious. I wrote this to try to fix in my own mind the extent of what I have done and to clarify for my own well being the probability that the experiment will be positive.

The flow diagram on the next page will perhaps help put the totality of the research into perspective. This flow chart depicts that I started with the three classical laws of thermodynamics and made deductions from them using restrictive assumptions. In the resulting conclusions I found all of the current theories, with perhaps some different interpretations, and some new predictions. In this chart E is the energy exchanged between the system and its surroundings. S is the entropy whether thermodynamic, mechanical, or both. P stands for pressure. F is a force. The Greek letter φ stands for a gauge potential component.

Of particular interest to me is the display of restrictive assumptions that replace additional assumptions in the standard model. For example, in the standard model to get Newtonian mechanics (NM) one needs to make three assumptions. To go from NM to the special theory of relativity (STR) another assumption (c=constant) must be made for a total of four fundamental assumptions. To go from NM to classical, or non-relativistic, quantum mechanics (QM) four additional assumptions must be made for a total of seven assumptions (3 NM plus 4 QM.). Maxwell's electromagnetism (EM) must be assumed in the standard model so this is an additional four assumptions. Here they are derived from the fundamental laws using Weyl's Gauge principle which is required by the laws. Gravity is an additional assumption in the standard model. Here it also is derived from the laws as the gauge function is expanded to five dimensions.

Why Physics?

1st. Law
$dE = dU + PdV - \overline{F}d\overline{x}$
2nd. Law
Fundamental Laws
Restrictive Assumptions
$\frac{dE}{\phi} \leq dS$:c=constant
Deductions
3rd. Law
Classical theories
$P \neq 0; \overline{F} = 0$
$P \neq 0; \overline{F} \neq 0$
$P = 0; \overline{F} \neq 0$
2D Classical Thermo
$dE = 0$
$dE \neq 0$
?
$dE \neq 0$
5D STR & E/M/G
$dS \neq 0$
STR & E/M
$dE = 0$
dS=0
dS=0
v<<c
$\int \phi_j dx^j = 2\pi i N$
5D Quantum Cond.
CoM
$\int \phi_j dx^j = 2\pi i N$
Quantum Cond.
Newtonian Mechanics
$\varphi_0 = \overline{\varphi} = 0; \varphi_m \neq 0$
Potentials Given
neutrinos
$\varphi_0 \neq 0; \overline{\varphi} = 0$
Conservation of Mass
$\varphi_0 = N_0 \dfrac{ke^{\frac{-\lambda}{r}}}{r}$
$\varphi_0 = 0; \overline{\varphi} \neq 0$
Strong Force
Phat photons
Quantum Solar Orbits Dark Matter Dark Energy
$\phi_m = N_M \dfrac{-GmM(1-bt)e^{\frac{-\lambda}{r}}}{r}$
3 Particles
Quantum Mechanic
2 Particles
Unlike Particles
Weak Force
GTR on 4D Surface
Unlike Particles Atomic separations
Atoms

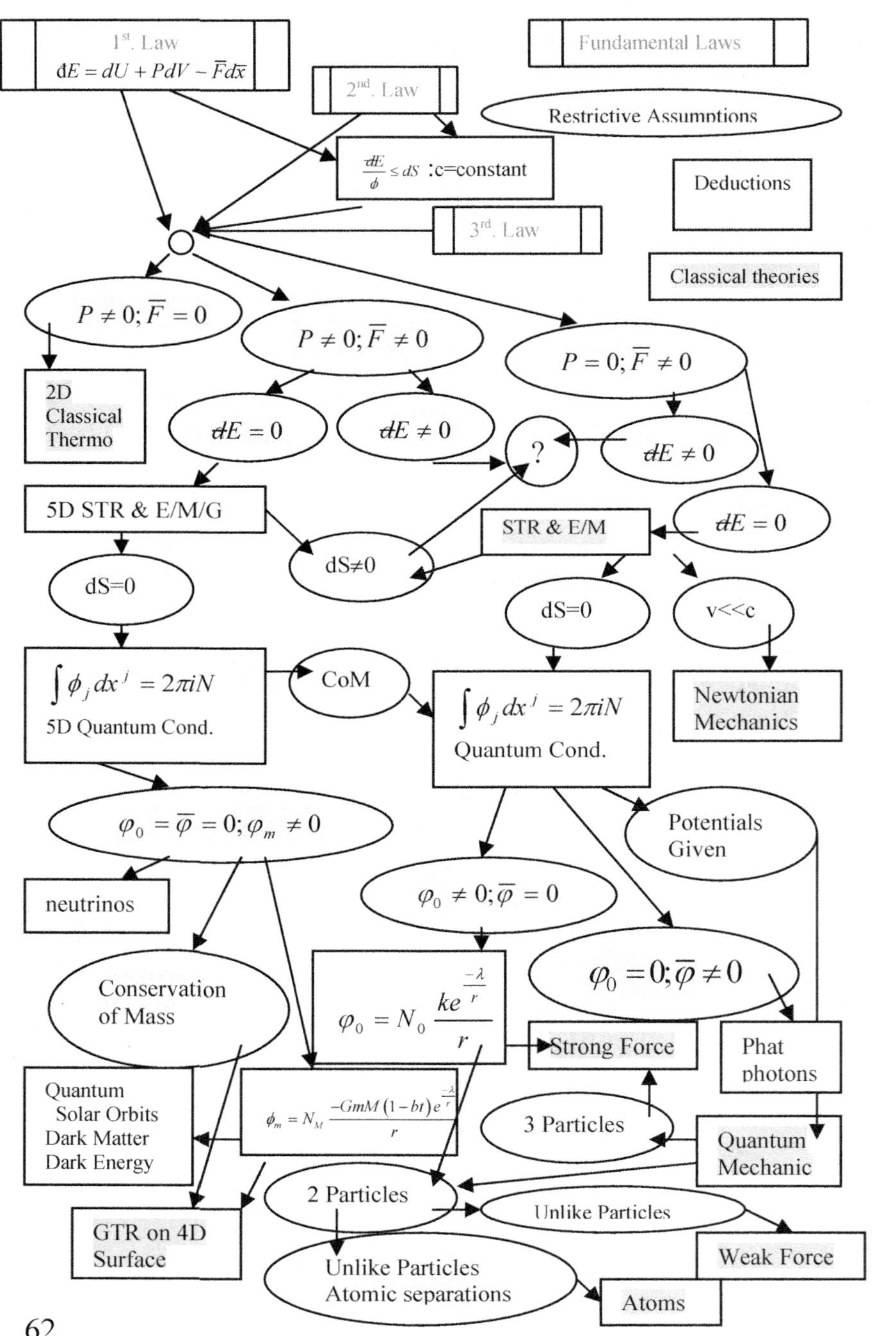

The value of this reduction in assumptions may be seen when the standard model is used in the atom. To obtain the Schrödinger wave equation with EM one must have NM (3 assumptions) plus QM (4 assumptions) and EM (4 assumptions). Each of these eleven assumptions is independent of the others and fundamental, meaning they cannot be obtained from any set of the other assumptions. Here we only need the fundamental three laws and a set of restrictive assumptions. Further, to get from the fundamental laws to QM with EM potentials only takes three restrictive assumptions; an isolated system, mechanical forces without pressure, and constant entropy.

Looking back over this development reveals several outstanding points. These include:

1. The Weyl Gauge Principle is at the heart of the development all of current physics as presented here and is thereby shown to have a fundamental role in four forces of nature, the electromagnetic force, the weak nuclear force, the strong nuclear force and the gravitational force. Indeed it was shown that all four forces stem from a single gauge field of five dimensions.

2. The simple restriction to phenomena for which Weyl's scale factor remains unity restricts the physics of gauge fields to those phenomena obeying quantum mechanics;

3. Immediately upon seeing that a unity Weyl scale factor requires quantization it was shown that new physics is predicted by this requirement. This new physics lies in the prediction of phat photons;

4. Once phat photons are predicted an experimental test of this new physics is seen

in the prediction of a phat laser. Once it is experimentally shown that phat photons exist the fundamental nature and validity of the Weyl Quantum Principle (WQP) is established;

5. After showing that the unity scale factor requires quantized energy of light when the vector potentials are involved, it was then shown—by looking at the electrostatic potential—that the Weyl Quantum Principle requires electric charge to be quantized in only unit quantities;

6. Further, the WQP requires that the electrostatic potential have a non-singular dependence on space such that it never becomes infinite and, therefore, does not need the renormalization necessary in classical quantum mechanics;

7. The non-singular potential leads to forces that violate Newton's third law of motion when two unlike particles interact and the equations describing these interactions were shown to be those of the Yang-Mills theory of the weak nuclear forces;

8. When three particle systems were considered the equations were shown to be of the form currently used for describing the strong nuclear forces;

9. The non-singular electrostatic potential produced the whole of atomic and nuclear physics and a five dimensional gauge field was shown to produce Einstein's gravitational field equations when the fifth dimension (mass density) was conserved;

10. Consistent units in a five dimensional gauge field requires a conversion factor involving a charge to mass ratio and this conversion factor was given by $\beta = \sqrt{\varepsilon G}$;

11. A five dimensional gauge field that inductively couples the gravitational field to the electromagnetic field provides several ways for these fields to produce effects normally associated with another field. Thus, a rotating gravitational mass will have a magnetic moment without an overall electric charge. The predicted magnetic moment of the Earth using this feature of the five dimensional gauge field provides a good prediction of the experimentally measured magnetic moment;

12. The cosmology determined using the non-singular gravitational potential requires changing current interpretations such as the big bang interpretation of the universe originating from a singularity. However, the exponential form of the gravitational potential provides theoretical support for inflationary cosmological models; and

13. Finally, the WQP was shown to produce quantum gravity and the quantized planetary orbits.

Therefore, the Weyl Gauge Principle (WGP) produces all of the four forces of nature and provides the unification of forces that has been sought for many decades. The unity scale factor restriction on the WGP reproduces quantum physics showing that it is a subset of the physics given by the WGP. Further, it is seen that the fundamental aspect of the restricted WQP may be experimentally proven by testing the prediction of phat photons.

Why?

One of the things I told myself when I began this study of physics was that by looking into the foundations of physics I would learn it better than I would if I did not. As you might have suspected from the foregoing at each step I spent a lot of time thoroughly understanding the fundamentals of current physics in the new field so as to best know how to proceed. This meant that I not only learned current physics, but how the new theory addressed the point as well. Therefore, I think I did learn more physics more thoroughly than I would have had I not been so interested in the foundations of physics.

Also, at each step I found agreement between the new approach and experimental data. I have not yet found disagreement between theory and experiment. I found new ways of looking at old physics as well as new physics.

I could not NOT do physics this way.

Chapter 2 Thermodynamic Basis

Fundamental Laws

Thermodynamics (TD) is probably the least liked subject of all the subjects that a scientist or engineer must take as they prepare for their careers. I didn't like it either. I can't really say I like it now. It has always seemed too vague or never within my grasp. Something has always seemed out of my reach. When I first ran into TD in college it was in a summer course and I thought it was a reduction in pain that I only had to sit in the class for ten weeks instead of 16. I escaped the class with a D. That didn't boost up my grade point average, but it did mean that I didn't have to retake it.

In postgraduate school I got my second dose of TD and I did a little better by making a B+ this time around. I suppose I had to do better as students in postgraduate school are supposed to maintain higher grade points. I can't say I liked TD any better the second time around, but perhaps I learned a little more about its vagueness. Or maybe I had learned that it was vaguer than I first imagined.

A short time after I took TD the second time I concluded, because of my search through the foundations of physics that I needed to get to know TD better. However, what I needed was to remove a little of the vagueness. I was looking for something in TD that apparently no one ever had searched for there.

When I first ran into Einstein's work I didn't like his special relativity theory either. At this far removed time I am not sure that I can correctly state precisely why I didn't like it, only that I felt that I didn't like something about it. As I began my postgraduate studies I started looking at the foundations of all the different branches of physics to see how well founded they were and if I might identify any points of weakness in them, special relativity began to change in my mind. This change was not due so much to my understanding relativity better, though undoubtedly I did, but due to a different picture of physics that was forming in my mind. The different picture forming in my mind involved a switching of the roles of forces and energy. By this time I had spent a few years on ships and had seen the strain that was placed on the lines tying a ship up to a pier in a river. The force of the water running into the ship could be pretty strong and of course depended upon both the size of the ship and the velocity of the current. If the lines were untied and the ship was set free it didn't speed up forever. The ship accelerated until it was zipping along with the velocity of the current.

What is relative here? In special relativity we are told that as the speed of an object increases its mass also increases. The mass increases until the force can no longer accelerate it to any greater speed. When the speed settles down it was found to be the speed of light. It bothered me to think about the mass increasing to infinity. For example, if the mass of the ship that was set adrift increased until the force of the water hitting it could no longer accelerate it would the ship sink due to its increased weight? After all Einstein also assumed that inertial mass was equal to gravitational mass. It doesn't take too much thought to reach the conclusion that perhaps the ship stopped accelerating because as it got faster and faster its velocity was closer to the current velocity and the force became less. When the ship had reached the velocity of the current there was no more force left. This argues that the force decreases as the ship's velocity increases. This would have no effect on the mass of the ship so now the ship at least wouldn't sink.

A difference in the philosophy of who is in charge began to form in my mind. Consider it this way. I was taught that force acting through a distance gave me energy. This is the concept of potential energy. If you raise a weight up in the gravitational field and hold it there it has the potential of doing some work if you let it drop down again. This ability to possibly do work is the potential energy. The same text book that I learned this fact of nature from also told me that force equaled the mass multiplied by the acceleration. That statement implies force is the driver. Force causes things to accelerate. It doesn't say much about where the force came from. Newton came up with a separate law that tells us that the force due to gravity is the product of the two masses involved divided by the square of the distance between them. Now suppose we let the forces of gravity accelerate our masses. Einstein's relativity tells us that the masses will increase until they become infinite as the velocity of the masses reach the speed of light. My mind begins to scream. What in the world does the speed of light have to do with gravity?

Now perhaps the difference of who is in charge may be seen a little better. I have no problem with the limiting speed being the speed of light for gravitational force if there is a fundamental connection between light and gravity. I even played with the relativistic equations of motion until I was convinced that we might just as easily argue that the force decreased to zero as the velocity approached the speed of light as to say the mass increased so much the force couldn't push it around. Now let's go back to the ship. I much prefer to think the force got less as the speed of the ship got closer to the speed of the water. It matches what I think of as common sense to think that the less differences between the speed of the ship and the current the less force. So were does the force on the ship come from? It probably comes from the potential energy in the river water. The water is trying to go down river due to the Earth's gravity pulling on it. This energy ended up causing the force because the ship wasn't going as fast as the water.

OK. If energy causes the force, how fast can the energy go? For the force on the ship in the river the answer would be the velocity of the current. For electromagnetic forces the answer would be the speed of light because that is the speed of the electromagnetic energy. If we also say that the speed of light limits the gravitational force as well we are implying some fundamental connection between electromagnetic and gravitational forces that had not been suggested until Einstein. By this perhaps you may see how my philosophical view of forces and energy turned around. Instead of thinking that forces create energy I began to think energy creates forces and if energy cannot travel from one place to another at an infinite velocity than there would be no way for it to create a force that could accelerate anything faster than the energy can go.

So now we reach the point where instead of force being in charge and equations of motion are statements about forces moving things, we reach the point where we need to learn how energy creates the forces that we must put into the equations of motion. The first law of thermodynamics is the ultimate statement of the conservation of energy. Can such a law tell how energy can create a force?

The other major theory in physics that has enjoyed tremendous success is quantum mechanics. It is the only theory that could address atomic physics. It is the only theory with which Einstein had problems accepting even though his ideas helped start it. What was it about quantum mechanics that bothered Einstein? We have all heard or read about his comment that God doesn't play with dice. But I don't think that was the main thing that bothered Einstein about quantum mechanics. I believe he was bothered by something similar to what bothered me about it. We all know that a string, a wire, or a rope will oscillate in a periodic fashion if tied down at both ends and given some energy. This is what makes the music that comes from all stringed instruments. These periodic motions are standing waves. Standing waves are generated in several realms of physics and cause me no problem. I even happen to like them, especially when in musical instruments. What I don't like about quantum mechanics is that the majority of the physicists interpret the success of quantum mechanics in the microscopic world to claim that means quantum mechanics is a more fundamental theory than other theories. This is as if describing small things should be more fundamental than describing large things. I think this was Einstein's objection to quantum mechanics. I don't think he objected to its success in atomic physics, but rather I think he objected to the claims of a fundamental character for the theory.

I know quantum mechanics is needed to help describe certain realms of physics. I even happen to find some of its predictions interesting, intriguing and exciting. I just don't think quantum mechanics is the end all theory it is claimed to be. I think it must be used when certain restrictions are placed upon nature. I just didn't know what these restrictions were. To be sure if we are to find that TD produces equations of motion we need also to find that quantum mechanics can be found within TD.

First Law: Conservation of Energy

The first law of TD tells us that some energy may be exchanged between a system and its surroundings that cannot be expressed as being due to the work the system is doing. It also equates this exchanged energy with the difference of any change in system's energy and work the system might do. Another way of expressing the first law might be to say that it lumps any left over energy after one accounts for all changes in the system's energy and the work the system has done into a single pile and the pile is

71

called exchange energy. In the case of a purely TD system this pile of exchange energy is called heat.

It was displayed long ago that TD heat and mechanical heat are essentially the same thing and, therefore, we now write the first law as an equation of the form

$$đQ = dU - PdV - F_x dx - F_y dy - F_z dz .\qquad (3)$$

Here the đ in front of the Q says that this quantity depends upon the path, or the changes of the system. The first term on the right hand side is the change in the system's energy and since the d does not have a bar in it we know this change does not depend upon the path. The next four terms involve the path directly as the elements of changes in the path are expressed as the dV, dx, dy, and dz. If there is only a dV term it is a purely TD system. If the $dV=0$ is the only vanishing work term it is a purely mechanical system.

There are a lot of things we can learn about any system using the First Law. By the same token there is a lot we cannot learn. For example, there are a lot of ways that the First Law can be satisfied and yet these are never seen in nature. Further, if we are looking for equations of motion that are to tell us how things move, the First Law cannot help us as it says nothing about which path nature may prefer.

There is one thing about the First Law that bothered me in mechanics classes. For simplicity let us think of setting all work terms to zero except one. Now our First Law has only two terms on the right hand side. Let us further reduce the problem by requiring that our system be isolated so that the exchange energy must be zero. Now we have a very simple looking equation stating the First Law for our restricted system. Let us also assume that the system's energy is due only to the energy of its motion. This equation looks like this

$$0 = mvdv - Fdx\qquad (4)$$

which can be written as

72

$$mv\frac{dv}{dt} = F\frac{dx}{dt} \qquad (5)$$

where I have moved the first term on the right hand side over to the left hand side and divided both side by an increment of time. We may recognize that the change in distance with time is the velocity and the change of velocity with time is the acceleration and our equation for our restricted system reads

$$ma = F \qquad (6)$$

after we have divided both sides by the velocity. Before we congratulate ourselves at arriving at Newton's statement that mass times the acceleration equals the force, I must point out that our First Law in Equation (3) above requires that the force must be velocity dependent and, therefore, the simple looking force in Equation (6) is NOT the Newtonian force that must be only a function of position like his force of gravity that depends only upon the separation between two masses.

All the text books on mechanics tell us that the Newtonian force of gravity is a 'conservative' force when used in an equation like Equation (6). We arrived at this equation from the First Law which I told you was the conservation of energy law. This sounds like double talk, and perhaps it is. It is definitely a point that we need to look into a little further in order to see that there are more than one way in which we speak of the conservation of energy and if we are not clear in pointing out the restrictions that are to be placed upon our system we will not be clear about what sort of conservation we are talking about.

Suppose I told you that the force appearing in Equation (6) was a function of velocity times Newton's force, that was only a function of position, then we might write the equation as

$$ma = F = \sqrt{1 - \frac{v^2}{c^2}}\,F(x) \qquad (7)$$

where we now see the force has been written as the product of two terms. One term depends only upon the velocity and the second term depends only upon the position. Now this velocity dependent force, the product of both terms, does satisfy our First Law with its requirement that the conservation of energy be path dependent. But the approximation that we might be tempted to make for slow moving systems, where the velocity is much smaller than the speed of light, does not satisfy the First Law because the approximation that drops the velocity dependence also drops the path dependence. However, the approximate equation that does not satisfy our path dependent First Law will satisfy a path independent relation that is called the conservation of energy in all the texts on mechanics. Note that this conservation of energy is a low velocity approximation to the path dependent First Law, but it should not to be confused with the First Law.

The reason I say the approximation is not to be confused with the First Law stems primarily from the fact that given the First Law and the velocity dependent force we may readily see that the low velocity limit is the path independent statement of energy conservation used in the texts on mechanics. However, the reverse is not so obvious. If I just gave you the simple statement of Newton's force that does not depend upon velocity you could deduce the path independent conservation of energy, but you would be lost in trying to deduce the First Law as you might not suspect that the First Law required the force to be velocity dependent and even if you suspected the velocity dependence you would then need to guess the exact function of velocity in order to arrive at the First Law. Einstein used his assumption about the constancy of the speed of light to deduce a modification to Newton's force, but did not indicate in his writings that he saw the implication about a different conservation of energy law.

Having said all this I now need to offer a solution to this apparent road block. Remember I said the First Law must be satisfied, but that it could not tell us which path to choose. The choice of paths must come from another law.

Second Law: Paths can't cross

The title of the Second Law that I chose to put in this second title may not be familiar to you. I chose it for that very reason. I might have used a title with the word entropy in it but that might conger up a notion of entropy that would be wrong. I might, and almost did, use the words free energy, but these also may not lend the necessary line of thought. So instead I chose to use the title "Paths can't cross" in order to point out this fundamental requirement of the Second Law and its development into statements about things like entropy and free energy.

When Caratheódory first stated his version of the second law of TD he did not say anything about paths crossing, which comes as a deduction from his statement. This is how Caratheódory stated his law:

> In the neighborhood (however close) of any equilibrium state of a system of any number of dynamic coordinates, there exist states that cannot be reached by reversible E-conservative $(\text{đ}E = 0)$ processes or motions.

This statement says there are points around any state of the system that can't be reached without exchanging energy between the system and its surroundings. One of the first conclusions that may be drawn from the Second Law is that solution paths for the First Law cannot cross. At first this may seem like a funny statement to find on the way to equations of motion, but it turns out that if we are going to find a statement about how to find the path that a system in nature will take it is likely to include something about paths that cannot be taken.

75

Saying that solution curves cannot cross turns out to do the job.

The next thing that can be determined from adding the Second Law to the First Law is that, because the paths cannot cross, there will always exist a mathematical function that when the First Law is multiplied by this function a new system property is formed that does not depend upon the path. Remember the First Law is a path dependent statement about energies balancing. This new system property was given the name of entropy for purely thermodynamic systems so I called this property mechanical entropy for purely mechanical systems. With continued usage it will probably be called simply entropy.

The word "entropy" is derived from the Greek εντροπία "a turning toward" (εν- "in" + τροπή "a turning"). In the modern scientific literature the word entropy is most commonly associated with the notions of disorder and chaos. However, in the study of boilers and large practical systems that combine heat and mechanical motion the concept most often called entropy is the concept of energy that becomes unavailable. We shall see as we go along that the other concepts of disorder and chaos will have their place also, they are just not as prominent as the current literature implies they should be.

The differences between the exchange energy and the entropy start with the multiplicative function that when it is multiplied by the change of the exchange energy changes the path dependent statement of the First Law into a path independent statement about the change in entropy. Mathematically such a function is called an integrating factor. It exists for the First Law only because of the Second Law and its requirement that paths cannot cross. It is a very powerful mathematical statement as it is a combination of both laws. We also see that each problem requires that we apply both laws and, therefore, each problem necessarily involves both exchange energy and entropy. Sometimes, but very seldom, the solution sought may ignore one or the other.

The Constancy of the Speed of Light

The existence of the integrating factor comes with an interesting requirement. The integrating factor is only a function of velocity! We haven't gotten far enough along to determine theoretically just what this function of velocity is, but we can quickly see that it is only a function of velocity. Actually we can say a little more. It is the reciprocal of a function of velocity only. So we should call our function of velocity an integrating denominator. It's not fair to peak back at the function of velocity I used earlier. We will learn what the function is in due time.

The next thing that we can do with our new found knowledge that the integrating denominator is only a function of velocity is to put it together with the fact that the change of exchange energy is path dependent while the change in entropy is path independent. The result of this is that we may arbitrarily establish a path. In this case suppose we chose any path that is traveled at a constant velocity. We are choosing a constant velocity path so that everyone whose coordinate system is moving with constant velocity with respect to our coordinate system will also say the path is a constant velocity path. Now we can form the ratio of the change in exchange energy at one velocity to the change in the exchange energy at some reference velocity that we may choose to call the velocity, c. We have chosen two paths, but we may also require that the change of entropy along both paths be identical. This means that when we form the ratio of the changes of exchange energy this also forms a ratio of the products of the integrating denominators and the change of entropy. By requiring the change of entropy for both paths to be identical the ratio of the change in entropy becomes unity and leaves us with the ratio of the change of exchange energies equal to the ratio of the integrating denominators. In this way we find that the change in exchange energy goes to zero as the function of velocity goes to zero. This means that as one approaches the velocity for which the function of velocity goes to zero no more energy can be exchanged between the system and its surroundings. Thus the velocity for which the function goes to zero becomes a limiting velocity. It is a unique limiting velocity because we have not been required to discuss any forces to reach this conclusion.

The unique velocity, required by the two laws, thus must be the same for all observers moving with a constant velocity with respect to our coordinate system. This is just what Einstein assumed when he stated his constancy of the speed of light. We didn't have to assume that the speed of light was a constant and the same for all observers moving with constant velocity with respect to each other; the two fundamental laws said so.

I did not dream up this route by which the two laws can be shown to produce Einstein's postulate of the speed of light. This trick has been used before in classical TD to show that there is an absolute zero temperature. The prediction of an absolute temperature has been borne out in many experiments through the years. Indeed the validity of the unique limiting velocity has also been borne out many times through the years. The fact that the two laws predict something so well proven in experiment in the field of mechanics argues that they do have dominion over the physics of mechanics. We might do as Einstein did and simply use the unique limiting velocity to modify Newton's laws, but that would not determine whether the two fundamental laws could provide them without the need to assume them.

Equations of Motion

The integrating denominator converts the path dependent change of exchange energy into a path independent change in mechanical entropy. This is the same statement found in TD where dividing the heat by the temperature produces the change in entropy. It is easy then to use Clausius' theorem for mechanical entropy as well as for the purely thermodynamic entropy. Clausius' theorem states that if we integrate the change in exchange energy divided by the integrating denominator over any closed path the result will be less than or equal to zero. The equality must hold if the path chosen is reversible. That is if the system can be taken backward along the same path and end up in the same state as it was when it left the integral would be zero. In this fashion one can prove that the integral of the change in exchange energy divided by the integrating denominator will be less than or equal to the change in entropy between the two end points. If it happens that there is no exchange of energy along this path the system is called an isolated system and for this system the entropy change must be greater than or equal to zero. This is the statement of the Entropy Principle.

To restate it: the Entropy Principle states that, for an isolated system, the path taken must be such that the entropy is maximized. This may be what we are looking for in our quest for equations of motion from the two fundamental laws, because this statement selects one specific path out of all possible paths between two points. That is the role of equations of motion. Previously, TD provided this statement and now we see that it also holds for mechanical systems as well.

While the laws state that an isolated system must maximize the entropy, we find that a non-isolated system must minimize a function of energy called free energy. As it turns out minimizing free energy for mechanical systems leads us away from our goal of showing that current theories come from the fundamental laws and, therefore, we will stick with isolated systems for the most part and leave non-isolated systems for a later discussion.

Geometry

Newton only had Euclidean geometry to work with and in using it he faced some criticism for using an absolute space and time. Einstein took a slightly different route and thought and showed that mass curved space. The non-Euclidean geometry that Einstein used is one in which the angles within a triangle do not add up to exactly 180 degrees, but the length of the vectors remain fixed. This was the only non-Euclidean geometry Einstein knew. Shortly after Einstein published his general theory Herman Weyl published an article that introduced a geometry in which the angles in a triangle did not add up to 180 degrees and the length of the vectors could change throughout the space so long as the length was guided by a certain function that Weyl called the gauge function. Weyl had introduced his new geometry in the hopes of unifying electromagnetic forces with Einstein's gravitational force in a curved space. Weyl thought that if the space curvature was used instead of a gravitational force then perhaps the changing length of the vectors could replace the electromagnetic force.

Weyl's hopes were quickly dashed by a criticism from Einstein that there was a path dependence showing up in Weyl's geometry that did not appear in nature. This criticism set Weyl's geometry on the sidelines. However, we will see it again below.

It was my belief that the fundamental laws should tell us what geometry to use. We should not be required to guess about what geometry to use. Therefore, I wanted to see if the two fundamental laws might specify the type of geometry we must use. So I looked at the different requirements necessary for equations of motion. There were two things needed to obtain equations that specify the motion taken by a system. First, one needs a variational principle. We have a variational principle even though we don't yet know how to use it. Next we need to determine if the laws could specify the geometry that was to be used.

In TD there exists a way to study of the stability of a system. This involves looking into just how stable an equilibrium state is. The procedure to investigate the stability is to look at how the system tends to move if it is displaced from the state under investigation. This investigation results in finding the stability conditions, or the conditions under which the system will try to return to a state if displaced from it. These stability conditions have a mathematical form called a quadratic form. This is the same form of relations needed to specify geometry so I began to investigate the stability conditions to see if I could find something that reduced to Einstein's special relativity because I knew if I could see special relativity then Newton's mechanics would also be present.

There are many ways to express the stability conditions. This involves using different sets of available variables. For example, velocity and energy might be chosen. The resulting stability conditions would be an accurate statement of the stability of the system, but they not only do not look like anything we have ever seen in mechanics, the resulting equations are next to impossible to solve. Of all the choices I found the one that led to mechanics as we now know it was to choose to write the stability conditions in terms of space and entropy. First, I needed entropy to be one of the variables because the Entropy Principle is the only variational principle available for use in our equations of motion. It turned out that combining space with entropy gave us the special relativistic equations. But we cannot see the special relativistic equations without going around an unexpected bend in the road.

As mentioned the stability conditions can be written in terms of space and entropy, but in doing so it leaves the distance between two points without a calibration. Something was needed to specify the distance between two points. Further, equations of motion give the change of positions in terms of a change in time. The stability conditions did not include any tie to time. However, every observer moving with constant velocity with respect to each other would be guaranteed of having the same limiting velocity so we can tie distance and time together by using this well-defined velocity which everyone will be able to measure. We can calibrate the stability conditions by setting the distance between two points to be the speed of light times the change in time.

Now we have an equation that gives the square of the speed of light times the square of the change of time set equal to a quadratic form involving the squares of the changes in space and entropy. Of course, as noted our Entropy Principle is stated in terms of entropy and not the time so this quadratic form must be turned inside out so we can write the change in entropy as being equal to something in order to use the only variational principle we have. This is the unexpected bend in the road.

Once the calibrated space-entropy stability conditions are turned inside out we find we have a new quadratic form that reduces to that of Einstein's special relativity with the change in entropy giving the distance between two points. But this isn't all. We find there are now two geometric manifolds to worry about. In the process of turning the stability conditions inside out we find a single multiplicative function ties the entropy manifold to another, energy manifold. Now instead of having only one manifold to investigate we have two. By using the facts that the entropy is the arc length of the entropy manifold and the Second Law requires that the entropy be integrable there is a straightforward, but tedious,

82

method of proving that the entropy manifold must have a Riemannian geometry while the energy manifold has a non-integrable Weyl geometry. The multiplicative function between the two manifolds is a geometrical integrating factor. The geometrical integrating factor is Weyl's old hypothesized gauge function!

When I discovered this I was really taken aback. First, the fundamental laws did indeed specify the geometry. Secondly, they specifed two manifolds and these two manifolds are two second order quadratic forms that mimic the two first order differential equations in the statement of the First Law and the mechanical entropy including the tie between them of the integrating factor. I had never read about a geometrical integrating factor and even Weyl's book introducing the gauge function did not talk about this role of the gauge function. But here it was; slick as a whistle. In TD we always had to see what the heat, energy, was doing and what the entropy was doing before the problem was solved. Here we see the same thing. There is a separate role for energy and entropy.

Arrow of Time

The fact that our equations of motion allow for time to run forward or backwards always has been a curiosity. Nature seems to desire that time only go forward. TD has always had its arrow of time in that for isolated systems the entropy seeks a maximum and for non-isolated systems free energy seeks a minimum. Here the same holds true for mechanical systems. The mechanical entropy is the arc length and the equations that maximize the arc length are the same equations that seek the shortest distance. But the mechanical entropy can never get smaller for isolated systems. If you compare the entropy manifold to Einstein's space-time manifold the entropy corresponds to Einstein's proper time.

The Second Law now states that for isolated systems the proper time can never run backwards. This is the same arrow that is displayed within TD.

Energy, Entropy and Forces

Energy is currently thought to be a derived quantity in our usual procedure within mechanics. In TD energy is the fundamental property of the First Law. Having shown that mechanics may be derived from the TD First Law we find that energy is the fundamental property of mechanics also. It is true enough that force is in the statement of the First law, but it takes the Second Law together with the first Law to arrive at equations of motion. Therefore, the concept of force does not play as fundamental a role as energy does.

The fact that the same procedures now used in TD can produce definitions for different energy concepts in mechanics shows that what has already been learned in one branch of physics may now be applied to another. Further, it may be instructive to write down the expressions for the various energy concepts that may be obtained by the above process for systems with only one mechanical force.

The only functional form for a force that fits the path dependence of the First Law is a force of the form

$$F = \sqrt{1 - \frac{v^2}{c^2}}\, \tilde{F}(x) = \sqrt{1 - \beta^2}\, \tilde{F}(x) \tag{8}$$

where the $\tilde{F}(x)$ means a function of position only and β is the ratio of the velocity to the speed of light. The force that is a function of position only is the force used in Newtonian mechanics while the function of velocity that shows up in the force is required by the path dependence of the First Law.

The system's energy in the First Law can be shown to be

$$U = \frac{m_0 c^2}{\sqrt{1 - \beta^2}}. \tag{9}$$

It should be noted that the system's energy is just Einstein's rest energy when the velocity goes to zero and becomes infinite as the velocity reaches the speed of light.

The entropy may be shown to be

$$S = \frac{\frac{1}{2} m_0 v^2}{(1 - \beta^2)} + V(x). \tag{10}$$

Here we see that the mechanical entropy is the sum of two terms. The first term is strictly a function of velocity and the second term is strictly a function of position. This suggests the first term be called the kinetic entropy; that is, entropy by virtue of having velocity. The second term should be called the potential entropy; or entropy due to position.

Chapter 3 Quantum Mechanics

Quantum mechanics (QM) has swept away all critics with tremendous successes in the atomic and subatomic realms. It is virtually stymied with respect to gravity though. Still quantum mechanics has had a much maligned past and still gets a lot of criticism and disbelief. Perhaps this disbelief is understandable given the different approach to obtaining solutions to motion problems that must be used in QM. Still most QM adherents claim that it is the most fundamental of all theories and the macroscopic theories can be derived from QM. I found it hard to believe this claim and felt that somehow QM came about when restrictions were placed upon the system under study.

Quantum Mechanics or Thermodynamics

It is commonly believed that one can "derive" thermodynamics from a variety of force laws using the techniques of statistical mechanics. This belief is not supported when one considers the development of statistical thermodynamics. For instance, in order to talk of a statistical temperature, τ, one must start by assuming Newtonian physics (this constitutes three fundamental assumptions). Given Newtonian physics one can talk of an energy distribution, canonical ensembles and statistical temperature; however, one must make an additional fundamental assumption (the Equipartition Law) before the statistical heat capacities may be obtained. In order to obtain thermodynamics we need two more assumptions. To display the assumption necessary for the first law of thermodynamics let me quote from page 85 of "Basic Theories of Physics: Heat and Quanta" by Peter G. Bergmann. "The difference between the heat transferred to the system and the work performed by it, is, according to our previous discussions, the increase in u. But in a systematically thermodynamic approach (that is, using only macroscopic observations and concepts), we get the differential expression, Eqn. (2.64) without reference to u. From that point of view, to claim that this expression is an exact differential is a logically new assertion; and this assertion constitutes the First Law of Thermodynamics."

The assumptions of statistical thermodynamics allow us to derive the differential of the heat exchanged using the statistical temperature and the statistical entropy. Further, it may be shown that the statistical expression for the second law of thermodynamics is analogous to the classical expression for an isolated system. In statistical thermodynamics it is asserted that the statistical temperature is related to the classical temperature by the Boltzmann's constant. Once this assertion is made then the statistical entropy is equal to the classical entropy. However, there is no logical necessity that the ratio of the statistical temperature to the classical temperature be a constant from the statistical approach, and only if it is a constant can we have a one-to-one correspondence between the statistical entropy and the classical thermodynamic entropy. The assertion that the ratio is a constant is logically equivalent to assuming the second law of classical thermodynamics.

The misconception that classical thermodynamics may be derived from Newtonian mechanics without the necessity of making additional assumptions is further entrenched by authors, such as Kittel, who in his text "Thermal Physics" says the following on p. 49, "We show in Chapter 8 that τ is proportional to the conventional absolute temperature which is measured in degrees Kelvin"; (This implies a logical necessity) On p. 427 the author states, "By analogy with the relation $dQ = \tau d\sigma$ we `assume' that the Kelvin temperature T has the property $dQ = k_B T d\sigma$ for a reversible process; here k_B is a constant to be determined and σ is the entropy." (The implied logical necessity is reduced to an assumption.)

Constant Entropy Requires Quantum Mechanics

In the previous chapter we introduced briefly Weyl's geometric proposal as an attempt to unify electromagnetism with Einstein's then recently published general theory. Einstein had sought to express the motion of a body under the influence of gravity as given by the path of minimum distance in a curved space. The geometry he used in his equations was Riemannian geometry wherein the angles in a triangle do not add up to 180 degrees. Yet in Riemannian geometry the length of a vector remains constant as the vector is moved around in the space. Weyl asked the question of whether or not electromagnetism could be described in a geometry where the length of the vector is allowed to vary as the vector is moved around in the space. He answered his own question with a very definite affirmative. He showed that the gauge function that controls the scale factor, and hence the length, of a vector was the origin of Maxwell's electromagnetism by deriving Maxwell's equations of electromagnetism from the gauge function. This may be called the Weyl Gauge Principle as Weyl himself did later in 1929.

The stigma of the necessity of having a path dependent metric, or geometry, doomed Weyl's proposal almost at the first published article. Einstein argued that Weyl's gauge function forced the metric to be path dependent. The gauge function led to the electromagnetic forces that held the atoms together. Therefore, he argued that Weyl's proposal would require the atoms to be dependent upon their histories and experiment did not show such a dependency. Therefore, Einstein concluded that Weyl's geometric extension did not correspond to nature and Weyl's proposal was dropped.

In the late 1910's and early 1920's the description of the atom was at the forefront of physics. It was by this time obvious that the orbits of the electrons around the proton in the hydrogen atom were dominated by the fact that the electron's orbital angular momentum was quantized. In 1922 Schrödinger noticed that these quantized orbits satisfied the condition that Weyl's scale factor was unity for every one of them. He thought this so remarkable that he published a paper to that effect. However, this remarkable feature of Weyl's unity scale factor was not mentioned in Schrödinger's paper wherein he published his wave equation that explained the atomic orbits and started QM on its journey of successes. A short time later London published a paper in which he wondered in print if the same Schrödinger that noted the remarkable feature of a unity Weyl scale factor was also the Schrödinger that published the wave equation.

In that paper London went on to show that if one started with the requirement that Weyl's scale factor had to be unity, and one was given the electrostatic potential, then the only paths allowed by these restrictions were the paths that were solutions of Schrödinger's wave equation. What London accomplished was to derive the Schrödinger wave equation from the requirement that Weyl's scale factor be equal to the value of one. This is, to me a very important result that was virtually ignored by the scientific community.

The restriction of the scale factor to unity was sufficient to require the basic equation of QM be the equation that specified the paths allowed. But what does this restriction mean? London went on to show that the scale factor and the wave function of the Schrödinger wave equation were proportional to each other. This result did not seem to register on the theoretical physics community that was in a feeding frenzy at the beginning of QM. Yet what is this saying? If, the interpretation of QM is correct

90

and the Schrödinger wave function is related to probabilities and it is proportional to Weyl scale factor that was restricted to unity, then the length of the vector, or the scale factor, is related to the probabilities described by the wave function. Isn't it interesting that the probability of something happening is the value one and the scale factor is required to be unity?

The question to be asked here is, "If QM can be derived from the requirement that Weyl's scale factor be unity, then can the two fundamental laws somehow require that Weyl's scale factor be set to unity for some reason?" If the laws require a unity scale factor for some reason then that reason is the reason for the existence of QM and can be used to derive QM from the two fundamental laws.

Suppose we return to the origins of the laws in classical thermodynamics. In TD a very stable system is an isolated system for which the entropy is a constant. An isolated system means there is no exchange of energy between the system and its surroundings. A constant entropy system is one whose entropy doesn't change. When we look at the energy and the entropy manifolds with the isolated and isentropic conditions we find that this is the requirement that the only remaining possible paths are those usually called null trajectories. Null trajectories are those paths that can be traveled without the arc length changing. Another way of thinking about this condition is that after traversing around a closed loop path the entropy of the stable, isolated and isentropic system must return to its original value. There are an infinite number of ways this can be done. First, the entropy can increase and then decrease back to its original value. Next it can go through a complete wave-like cycle of increasing, then decreasing to a lower value than it had at the origin and then returning back to its original value. It should be possible to now imagine the infinite number of ways that the isolated and isentropic condition can be met with different paths.

London referred to this situation by pointing out that what is waving in the Schrödinger wave equation was the scale factor.

Weyl's Quantum Principle

Weyl proposed his scale factor as a gauge principle from which electromagnetism may be derived. London's work showed that Schrödinger's equation could be derived from Weyl's scale factor given the requirement of a unity scale factor and an electrostatic potential. If we are given Weyl's gauge principle we may derive the electrostatic gauge potential from this principle and need not be given the electrostatic potential as stemming from any different principle. The only remaining condition that London used was the unity scale factor. In this case, since Weyl's gauge principle states no limitation on the scale factor in deriving electromagnetism, we now see that the isolated, isentropic system forces the restriction of a unity scale factor and propose the following as Weyl's Quantum Principle:

> A space that satisfies Weyl's gauge principle and for which Weyl's scale factor is unity, is said to obey Weyl's Quantum Principle.

With the statement of a single principle, Weyl's Quantum Principle, we have the foundation from, and the guidance by, which we may derive both Maxwell's electromagnetism and Schrödinger's quantum mechanics. Therefore, Weyl's Quantum Principle is the only fundamental principle we need to derive Maxwell's electromagnetism and Schrödinger's quantum mechanics. Schrödinger needed more than this to develop his wave equation.

In the development of his wave equation Schrödinger used several things in a plausibility argument, but this argument can not be called a derivation such as London produced in 1926. Rather, he knew that a wave equation for a stretched string could be derived from

92

Newton's laws and he knew that whatever he arrived at had to be consistent with Einstein's photon energy relation and a postulate by de Broglie that related the wavelength of matter waves to the momentum. He also required that the wave equation to be linear in the sense that a combination of any two solutions would also be a solution and he required forces that were the negative gradient of a potential. Thus, on top of Newton's mechanics he required conditions from the foundations of quantum mechanics.

What Schrodinger achieved was a marvelous accomplishment and a tremendous success! What the Copenhagen interpretation of the results did was to mislead physicists for nearly a hundred years. If you look in almost any book on the basics of QM today where the development of the Schrödinger equation is presented you may find the admission that the development does not have the strength of a derivation, or that it lacks logical necessity, but you will also invariably find words to the effect that 'we could not expect to be able to derive the quantum mechanical wave equation from any of the equations of classical physics." This attitude by the big names of the time reduced the desire of anyone to look into the possibility of doing something like London did, or ignoring it when London did. The Copenhagen interpretation also led to the attitude that if QM can't answer the question the question should not be asked. This is an attitude that is so different from the heart of science that it still seems hard for me to believe scientists really have believed this position all these years.

Unit of Action-Uncertainty

Current interpretation of the uncertainty principle basically is that of a universal unit of action. The uncertainty principle has two parts. The first part states that position and momentum cannot simultaneously be measured to the exact values of the position and momentum. Rather, there is an inherent limit to the certainty that can exist in our simultaneous knowledge of position and momentum. The second part of the uncertainty states a similar limitation of the simultaneous knowledge of energy and time.

The manner in which this uncertainty enters into the solution of problems in QM is through a mathematical concept called the quantum Poisson brackets. At this point the important thing to note is not the actual detailed procedure of how to construct the Poisson brackets only that they must be used and that geometry plays a role in the resulting unit of action that comes from the quantum Poisson brackets. The way that geometry enters into the unit of action is through the differentiation required by the brackets. The Poisson brackets look into how certain functions vary with respect to position and momentum. This requires geometric differentiation. This means that the geometry plays a role in the answer. In fairness to the scientific community it must be admitted that there was little reason for this to be investigated until now. For example, when Einstein showed that gravity caused a curvature in space-time the scientists may have looked into the effect of this geometry on the results of the quantum Poisson bracket where they would have found that within the space occupied by an atom the curvature would not have had any appreciable effect on the resulting unit of action. Therefore, a Riemannian manifold, curved by gravitating mass, would not change the unit of action or the uncertainty relations measurably in atomic physics.

A manifold wherein the scale factor varies with a gauge function does produce a change in the unit of action. When both the classical and the quantum Poisson brackets are formed in a Weyl geometry the gauge function enters into the results and the gauge function then controls the unit of action. The basic interpretation of the uncertainty principle with respect to the simultaneous measurement

94

accuracy of position and momentum or energy and time remains qualitatively the same, but the gauge function changes the limiting uncertainty. We will see this influence of the gauge function in several places.

Non-Singular Potential

Returning again to London's work we may note that Weyl's scale factor depends upon a path dependent integral called a line integral. Basically a line integral is made up of three parts. The first is a function of the independent variables which may be called the integrand. Next there are differentials of the variables. These indicate small steps are to be taken along some path. The third part is the answer. In London's work he assumed he was given an electrostatic potential which is the integrand and required the answer to be given. London then asked what paths were possible if the electrostatic potential is given and the answer had to be such that the scale factor was unity. The answer he found was that the paths had to be those that satisfied the Schrödinger wave equation.

Usually when a line integral is found in a mathematics text the integrand and path are given and one is asked to find the answer. London turned the question around slightly. Yet another twist to the problem can be posed. The twist involves seeking to know why it is that the electric charge is always found to be quantized in unit steps of electric charge. At least the particles that we can measure in the lab have only unit charges. Later, we will address those hypothetical and elusive particles given the colorful names of quarks that are thought to have charges of 1/3 of a unit.

The electron and the proton are very stable particles. They last a long time and retain their properties of mass and electric charge in spite of us moving them around at high energies and velocities. In TD we remember that a very stable system is an isentropic system and we have already seen, from London's work, that an isentropic mechanical system acting under the influence of an electrostatic potential must have quantized paths. The isentropic condition is the requirement that imposes the Weyl Quantum Principle (WQP). Can the WQP specify that electric charge must be quantized? Suppose we wish to look at a fundamental particle such as an electron whose properties do not depend upon the path. The property of a charged particle is given by its electrostatic potential. Now we may state the next question that may be asked of the WQP. We shall require an isentropic particle whose electrostatic potential does not depend upon the path. This means we specify the line integral's answer and path and seek to determine what properties the electrostatic potential must have to satisfy the WQP. The answer has two pieces. The first piece is rewarding in that the electrostatic potential must be quantized. This, of course, is a well-known experimental fact of nature, but this is the first time it has been shown to be required by some fundamental theoretical requirement.

The second piece of the answer is the most surprising and ultimately the most rewarding. The second part of the answer is that the quantization of the WQP adds a multiplicative term in the electrostatic potential. At large distances, but still less than atomic separations, the new quantized electrostatic potential is essentially the old familiar electrostatic potential we get when solving Maxwell's equation with continuous functions. Notice that I specified a solution "with continuous functions" as I will return to this shortly. At sub-atomic distances the new quantized potential begins to turn back toward zero as the distance from the charge approaches zero. This behavior is called non-singular where the old electrostatic potential tended toward infinity as the distance went to zero in behavior that is called singular.

I ultimately found that this non-singular behavior of the electrostatic gauge potential extends to gravity as well, but that is another story to be told later.

Mathematicians have long known that continuous functions behave differently than do discontinuous functions. Many mathematical theorems specify that the theorem holds only for continuous functions. Maxwell's equations of electromagnetism are partial differential equations that are used to describe the electromagnetic fields whether these fields be waves or other types of fields. Usually when scientists and engineers use Maxwell's equations to find some solution they seek a solution of continuous functions. When this approach is taken one finds the static solution for a free charge is the Coulomb potential that decreases inversely with the distance from the charge. The potential allowed by the WQP is quantized and, therefore, is in the realm of discontinuous functions. This quantization, or discontinuity, of the potential adds the term to the solution that causes the potential value to return to zero as the distance from the charge goes to zero. The non-singular potential is the same one displayed in Equation (1) and is restated here

$$\phi_0 = \frac{Z_1 Z_2 e^2}{4\pi\varepsilon_o r} e^{\frac{-\lambda}{r}} . \tag{11}$$

The WQP requires that one must use QM if given a gauge potential and you wish to know the isentropic paths of particles interacting with this potential. On the other hand, the WQP quantizes the charge if you seek to know what gauge potential can exist that is independent of where or when the particle exists. In addition to requiring that the potential be quantized, the WQP also requires the potential to be non-singular for this very stable particle. These are new theoretical requirements never seen before.

One thing to remember that I have not pursued is that if you seek to know what electrostatic potential exists for a particle that does not satisfy the isentropic condition, it will only have a singular potential. I haven't yet had the time to look into such a particle, but I feel it too must exist. I just don't know what it is, or what they are.

Phat Photons

Einstein never did feel that the theory was complete with respect to the wave and particle view of light. The wave-particle duality discussion exists still. The Copenhagen interpretation is that the wave and particle descriptions are complementary. Einstein felt this did not answer the deeper theoretical foundation that required this duality. He still thought the there must exist a more fundamental theory in which the wave and particle aspects of the light would be required.

I related earlier that I had asked for help from God on different occasions. This was one of those occasions. I wished to know if the theory that was unfolding in my investigations yielded any illuminating statements with regards to the wave-particle duality. After my request for help I immediately thought to myself, "What does the WQP have to say about the gauge vector potentials?" The answer is rather easy to obtain after seeing how London used it in deriving QM and how it may be used to quantize the electrostatic potential. The vector potential is called the vector potential because is consists of three components making up the remainder of the gauge potentials from a four dimensional gauge function of space and time.

I supposed that one can always polarize light and, therefore, thought of all gauge potential components as being set to zero with the exception of a single vector potential component. I then sought to learn what an isentropic gauge vector potential component must be like. Then I found that this component must be quantized, but still it must satisfy Maxwell's wave equation. It must then be a quantized wave. It had all the wave properties that

98

Maxwell's wave equations produced except that they were quantized. When I looked at the energy by looking at the usual Poynting vector I found that the energy of the wave was quantized! This was just what Einstein had argued and indeed had used statistical arguments to show back in 1905. There was one difference. This difference may turn out to be a big difference in the science of the future.

Einstein derived the energy relation for his quanta of light from statistical methods and arrived at the expression

$$\varepsilon = h\nu. \tag{12}$$

The energy of the isentropic quantum of light that I derived was given by

$$\varepsilon = N^2 h\nu. \tag{13}$$

This expression may not at first glance appear to be that different from Einstein's, but my investigations showed that it leads to some very different physics that can prove to be of practical utility to man.

First, the appearance of the quantum number in the energy expression for the photon means that there may be photons of a given frequency that have more energy than the Einstein photon. So I call these photons, with N>1, phat photons. Secondly, the expression for the energy of the phat photons argues that an electron in a higher energy orbit around the proton in a hydrogen atom would have the possibility of emitting an Einstein photon or one of the phat photons. One of the first questions that came to me was "Have these phat frequencies been reported?"

So next I went through the mathematical procedure to calculate the probability of the phat frequencies under both spontaneous and stimulated conditions. The probability of emitting the phat frequencies diminish inversely as N^2 which means the N=2 frequency should be seen one fourth as often as the fundamental frequency and the N=3 frequency should be seen only one ninth as often. Thus, emboldened I went through the equations to determine the properties and probabilities of a phat laser. Such lasers should be possible.

Next I went to the National Institute of Standards and Technology web site and looked at the original data reported by C.E. Moore and there they were! The fundamental frequency for hydrogen was listed with a magnitude of 1,000, the N=2 frequency showed a magnitude of 80 and the N=3 frequency magnitude was 12. J. D. Garcia & J. E. Mack reported similar numbers for the helium atom.

What might be the benefits of using a phat laser? Consider that the phat frequencies represent greater energy at a given frequency for each phat photon. Therefore, in seeking a benefit of phat photons one needs to keep in mind what you are asking the photon to do. For example, suppose you wish to know how phat photons might be used in transmitting data across the country using fiber optics. The limit to the energy that can be passed through the fiber before the fiber begins to break down is controlled by the electric field strength of the photon because it is the electric field that interacts with the atoms of the fiber and causes them to break down. The energy of the phat photons is proportional to N^2, but the electric field is only proportional to N. This means that the electric field may be held at a fixed in value in the fiber if we reduce the number of photons by a factor of 1/N while we increase the frequency by a factor of N and, though the electric field strength is the same as it was the transmitted energy has been increased by a factor of N. We get more energy through the fiber optics by going to higher quantum numbers while keeping the fiber from being damaged by the additional energy. A secondary advantage in this example is that the number of required amplifier stations needed in a long transmission fiber will be reduced by a factor of N as well.

100

Chapter 4 Nuclear Forces

The current model of the nuclei is given by the standard model of nuclear physics which describes interactions of the weak and the strong nuclear forces. This model is primarily a phenomenological model and not a predictive theory. Yet it has enjoyed several successes. Plus, it has had almost no rivals.

In this chapter I shall present a very different approach to nuclear physics. This new approach will show how the non-singular electrostatic potential describes the interactions currently ascribed to both the weak and the strong nuclear forces. Therefore, suppose we again look at the basic properties of the non-singular potential. Its dependence upon distance was given as

$$\phi_0 = \frac{Z_1 Z_2 e^2}{4\pi\varepsilon_o r} e^{\frac{-\lambda_p}{r}} \tag{14}$$

where the subscript of λ_p indicates the fact that the lambda depends upon the particle and will be different for each type of particle.

The Strong Force Is Between Like Particles

The non-singular potential requires that the force between like particles be repulsive at large distances, but this force goes to zero when $r = \lambda_p$ and then becomes an attractive force as the distance decreases from λ_p on down to zero. This feature of being repulsive at long range and then becoming attractive at very close range is very different from the singular Coulomb potential. As a matter of fact it is so different that my first thought was that this could not possibly correspond to what we see in nature. Yet, I then remembered that is exactly what we see in nature.

When Rutherford first started allowing alpha particles to strike metal foils he learned a lot of new information about how small the sizes of the nuclei were and how much space was in between them even in material that seems so solid. But he also saw the beginning of the effects of firing one particle at another. This procedure is, of course, in the realm of high energy physics as done today. The latest and greatest machine to create collisions between particles is the Large Hadron Collider (LHC) at CERN. The LHC also seeks to study the collisions of very high energy protons.

The results of all the proton-proton scattering, starting right at the beginning, has supported the concept that at long range protons do not like other protons. At short range they change their effect on each other and become attracted to each other. When this feature of nature was first discovered it placed the scientific community into a position of needing the make a decision about such a drastic difference between the predictions of the singular coulomb potential and the experimental results.

There were not a lot of choices available to the scientists. On the one hand they had Maxwell's equations that gave them the Coulombic potential that led to the prediction of completely repulsive behavior between protons and from experiment they had this short range attractive behavior. Maxwell's electromagnetism was even at that time a very well verified theory and, therefore, it would be hard to consider the need to modify it. It would probably be even harder to learn what modification could be made that would allow electromagnetism to continue to explain what it was already known to explain and still account for this strange new attractive character. Therefore, they concluded that this new short range attractive behavior must come from some new force that had not been seen before. This new force had to be independent of electromagnetism so that a modification of Maxwell's equations would not be necessary.

The new force worked within the nucleus and had to be very strong to overpower the Coulomb force as it did so it was called the strong nuclear force. Another force was found to operate within the nucleus, but it was weaker than this short range force between protons so the other short range force was called the weak nuclear force.

I knew of these forces when I derived the non-singular potential and one of the first things I did was to derive the relations that told how the non-singular character of the new potential changed the predictions of proton-proton scattering events. Of course, as you may have already suspected, the results indicated that when two protons approached within a separation of lambda the force began to weaken rapidly. At a separation equal to lambda the force became zero and then was attractive at closer separations. (See Figure 4.1) This is just the effect seen in the scattering experiments. Plus, it came from the non-singular character of the new potential and this new potential also came from Maxwell's electromagnetism so a new force was not needed!

The scattering data supported the predictions of the non-singular potential and showed that the value of the lambda of the proton was of the order of 10^{-15} meter, or 1 Fermi. All scattering data from even the highest energy proton-proton collisions support this non-singular characteristic of the potential and the long range repulsive/short range attractive aspect of the force between two particles of like charges. At least it did for the protons.

What about the electron-electron scattering data, would it support the non-singular potential? The scattering data even at the highest energies currently used in electron colliders has not shown any tendency for the force to change from its long range repulsive character to a short range attractive character and the distances between electrons have been probed down to approximately 10^{-17} meters. This separation is much smaller than the separation where the proton began to display non-singular character so if the electron potential was also non-singular then the lambda for the electron had to be some three orders of magnitude smaller than the lambda of the proton. Further, because electrons have such a small mass the acceleration of electrons in an accelerator causes the electrons to emit photons and loose a lot of the energy that is trying to accelerate the electrons. For this reason it is very difficult, if not impossible, to accelerate electrons to higher energies than reached with the current electron accelerators. So for the moment let us simply say that if the electron potential is non-singular, its lambda is much smaller than the lambda of the proton.

The exponential terms in the non-singular potential make obtaining solutions to equations of motion with such terms next to impossible without the use of numerical simulations. I did not have the time nor access to someone who could write the computer code to solve the relativistic quantum mechanical equations in order to study the predictions of possible nuclear orbits of protons around protons. Also, a nucleus with a positive charge of two units and only two mass units did not seem to be a stable isotope of helium so an effort in this direction did not appear productive.

The Weak Force Is Between Un-like Particles

The non-singular potential requires that the lambda of different types of particles be different. The scattering data of protons and electrons show that there must be three orders of magnitude difference between their lambdas. This means the force between un-like particles violate Newton's Third Law. This conclusion throws a wrench into the theoretical works! Forces are not supposed to violate Newton's third law. Or at least that is what Newton thought and many scientists today think this is a law of nature that cannot be violated. However, there is one feature of Maxwell's electromagnetism that shows the forces between particles violate Newton's in certain specific conditions. Though this aspect of electromagnetism does not greatly influence our day-to-day living and is seldom discussed in the literature. Further, since I was trying to study the implications of the fundamental laws, I was reluctant to impose Newton's third law until I found that the two fundamental laws required Newton's third law to hold.

For several years I struggled with attempts to make approximations to the equations of motion in order to get a sense of what was predicted in terms of interactions of unlike particles. These attempts did display the general characteristics of the interaction, but it is best not to go into all of these and go straight to the best approach. This approach became very enlightening in the end.

One of the biggest advantages of having Newton's third law hold is that this law guarantees that a two-body problem can be solved as a one-body problem and, thereby, offers a great reduction in the work of solving the problem. This advantage alone is not sufficient justification to hang onto some law. If it is violated once in nature then it should be taken as a convenient law but not one that can never be violated. However, if Newton's third law is violated then any problem that involves two particles must be treated as a two-body problem.

One way of treating a two-body problem is to set up the equations to follow the motion of each particle. When I did this I could not readily see the motion collapse back to familiar motion when the forces obeyed Newton's third law. Since I knew what the motion should look like if Newton's law held, I wished to set the problem up so I could easily see if whatever I got collapsed to the known when the known was used. Then I decided to set up the problem in a fashion that I should have seen at the start. It is standard procedure to set up the problem of two bodies to calculate the motion about the center of mass which sets still when Newton's third law holds. This allows us to use the concept of a reduced mass that compensates for the combination of the mass of the two bodies and a vector that keeps track of the position of the center of mass.

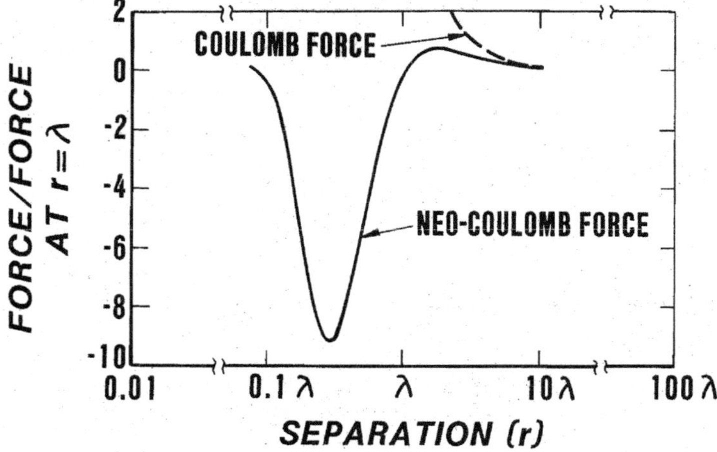

Figure 4.1 The non-singular vs. singular, coulomb force

I then used the motion of the center of mass and the motion about the center of mass instead of the motion of particle one and the motion of particle two. This way I had the two-body problem set up in such a way that I could easily see Newtonian motion. If I used forces that obeyed Newton's third law the center of mass motion stopped and the motion about the center of mass was just the familiar Newtonian motion. The need to keep track of two motions within the same problem of course caused an additional complication, but it made one value of using Newton's third law very obvious. Newton's third law really simplified the problem; however, it also gave the wrong answers if the force did not obey Newton's third law.

There were some interesting aspects of central forces that came to light when the problem is approached

106

assuming Newton's third law does not hold. A central force is one that only depends upon the distance the body is from a given point. Usually this point is the center of mass in Newtonian mechanics. Central forces have no component of force other than in the radial direction. This then requires that the angular momentum be a constant. When one studies a two-body problem set up to look at the motion of the center of mass and the motion about the center of mass, but the forces only depend upon the distance between the two particles, one soon sees central forces slightly differently. Of course, the center of mass must be on a line between the two masses. However, if the forces do not obey Newton's third law and the center of mass begins to move just a little it also must move around a point in space and this motion will have constant angular momentum. Further, if you have one large body and one very small body the center of mass is very close to the larger body. For forces that are only slightly different from Newtonian the center of mass motion may still move around a point that is close to the larger body. But this is not required. Some forces, such as the force given by the non-singular potential, can have the center of mass moving around a point in space closer to the smaller body than it is to the larger body. The aspect of central forces that requires the angular momentum to be a constant comes from a force that depends upon the distance between the two particles and is in each case directed along this line. Then not only must the angular momentum of the center of mass motion be a constant, so must the angular momentum of the motion about the center of mass.

When the solution of the problem is in hand the energy of the two body system is found to be the sum of four terms, the kinetic and potential energy of the center of mass motion plus the kinetic and potential energies of the motion about the center of mass. By setting up the problem in terms of the motion of, and about, the center of mass it is easy to look at the potential energy expressions in the problem solution set and to see Newtonian mechanics as the limiting condition when Newton's third law is satisfied. The same is true of the two expressions for the kinetic energy.

Neutron

What is the solution for the motion due to the non-singular electrostatic potentials of an electron and a proton? When the separation between the particles is much larger than the lambda of both particles the solution is for the electron to orbit around the proton in atomic sized orbits. But when the separation is of the order of the lambda of the proton (10^{-15} meters) the motion drastically changes. The electron feels almost no force while the proton still thinks the electron is a pretty cute little bugger and is still very attracted to it, so the motion is one where the proton orbits around the electron at a separation approximately equal to the lambda of the proton. See Figure 4.2.

Now doesn't that fry your bacon!

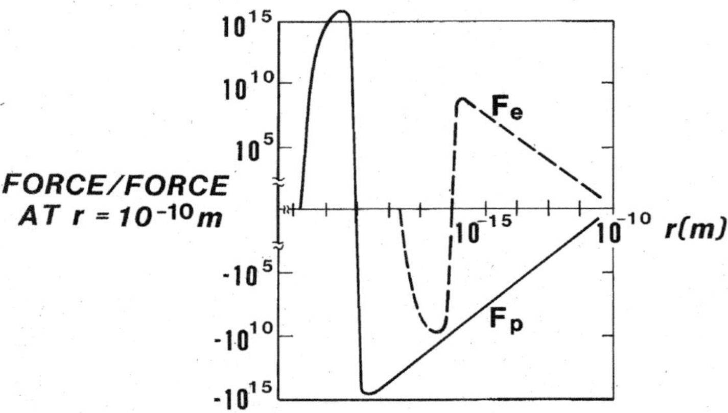

Figure 4.2 Unlike particle forces

There is another aspect of this motion that is important to note. In atomic orbits the electron is trapped in stable, negative energy states. However, these much smaller proton orbits are unstable because they are in positive energy wells. This means that where you must supply energy to the atomic states to pry the electron out of its orbit. The proton will come out of its positive energy orbits and have energy left over. (The positive energy well of the neutron is shown in figure 4.3.) Surely such behavior would be seen in nature.

The neutron displays just such a breakup. A neutron decays in roughly 10 minutes into a proton and an electron plus the center of mass moves. The motion of the center of mass is the energy left over when the proton and the electron tunnel out of the positive energy well. Now this is not what the standard model tells us about the decay of the neutron. The standard model tells us that, first, there is neither proton nor electron within a neutron, and second, the left over energy is carried away by a particle that was hypothesized to keep the center of mass of the decay products still. In this way the hypothesized neutrino reinstates Newton's third law.

The neutrino is not needed when the non-singular potential is used. Plus, the Wyle Gauge Principle (WGP) has something important to say about the neutron also. However, the proton orbit in a neutron should be relativistic since the proton should be going very fast in such small orbits. We should really save some of our discussion about nuclear orbits until we develop the relativistic quantum mechanical equations of motion and solve them.

As was mentioned earlier, Schrödinger assumed Newton's third law held when he developed his wave equation so his equation will not work for the non-singular potential on interactions between un-like particles such as the proton and the electron. I had to develop a new wave equation where I did not make the assumption that Newton's third law was to be enforced. This development followed the one I just described for non-relativistic, nonquantized systems. Indeed the fact that I had just gone through the non-quantum development helped greatly in

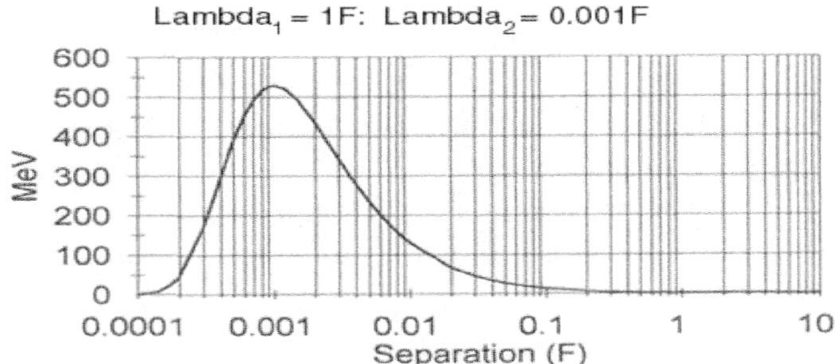

Figure 4.3 The Center of Mass potential.

the development of the revised Schrödinger wave equation.

I found that I could solve the resulting wave equation for the quantized orbits of motion of the center of mass and motion about the center of mass. The answers appeared more difficult as was to be expected, but they collapsed down to the standard model solutions for Newtonian forces. But this solution was not able to address the relativistic nature expected of the nuclear orbits, so something additional was needed. Before Dirac developed his relativistic quantum mechanics, there were relativistic approximations made to the solution to Schrödinger's wave equation. These approximations basically assumed relativistic velocities for the orbits and these approximations made important relativistic corrections to the solutions even though they were not the final answer.

So I made similar relativistic approximations to the solutions of the modified Schrödinger wave equation by the method of assuming relativistic circular orbits. The resulting solutions were very useful in comparing the predictions of the non-singular potentials to experimental data. The primary characteristics of the orbits remained the same, of course. The neutron was still one proton in orbit around one electron. This is not allowed in QM as currently interpreted. The primary reasons that this is not acceptable in the standard model is that the uncertainty relations are used to argue the electron cannot be confined to as small a volume as is occupied by the neutron and the only way to keep the center of mass from taking off is to assume that a neutrino is emitted as well as the proton and electron.

Earlier I pointed out that the unit of action depends upon the gauge function and for the gauge function values within the confines of the neutron the units of action for both the electron and the proton orbits are not given by Planck's constant. Another way of saying this is that the gauge function controls the particle's spins and the orbital spins. This shows up in the measured magnetic moments of the free electron, the free proton and the neutron where the magnetic moment depends upon the particles' spin and the orbital magnetic moment. The magnetic moment of the free particle is determined by one half of a unit of action, or one half of Planck's constant. The magnetic moment of the same particle in its orbit within the neutron is one half the effective unit of action for the orbit which will be some reduction of Planck's constant. One set of experimental data that is used in the standard model to argue the necessity of a neutrino emission is that spins must be conserved and the magnetic moments are the experimental indicators of spin. Does the standard model argument rule out the non-singular potential?

The equation stating the conservation of angular momentum and the accounting of the magnetic moments of the particles and orbits within the neutron that compares with the experimental magnetic moment provides two equations in two unknowns. The two unknowns are the effective units of action for the electron and the proton orbits within the neutron. These give the value of the effective unit of action for the proton orbit as $\hbar_p = 0.66586\hbar$ while the effective unit of action for the electron orbit is $\hbar_e = 8.0517x10^{-4}\hbar$. This says the conservation of spins and the experimental magnetic moments are satisfied by these values of effective units of action.

Next one can use the mass of the neutron to estimate the orbital radius of the proton. First, it may be noted that the lambda of the electron is much, much smaller than the lambda of the proton and can be approximated by setting the lambda of the electron to zero. This provides a radius of $r_p = 1.5x10^{-19}m = 1.5x10^{-4}$ *fermi*. We have now used three equations to find three unknowns. Essentially, we have not predicted anything. We have only shown consistency with experimental data. There must be some way to distinguish between the standard model and the non-singular potential.

The standard model has no means by which to predict when a neutron will decay. The non-singular potential provides the functional form for the positive energy well within which the neutron center of mass is trapped. This allows the use of the procedure of quantum tunneling to calculate the half life of the neutron. Quantum tunneling uses the energy of the neutron's decay products, the orbital velocity of the center of mass, the size of the center of mass orbit and the functional form of the positive energy well to predict the probable rate of decay for neutrons. The values of the orbital parameters determined above predicts a neutron half life of slightly over 600 seconds which compares favorably with the experimental value.

Deuterium nucleus

Once the relativistic equations have been solved for a single proton in orbit around an electron it becomes rather easy to include a second proton in the equations. When this is done there is symmetry to the proton orbits since when

112

protons are at a separation larger than something like a Fermi they hate each other so it would take some sweetening to get two protons to get close to each other. In this case the sweetening is the electron. Recall that when the separation between an electron and proton is of the order of the lambda of the proton, around a Fermi, the electron tends to ignore the proton. But certainly the electron is the glue that gets the two protons to share space within this nucleus.

Two protons and a single electron make a nucleus with a single positive charge unit and two mass units so this is a deuterium nucleus. The standard model tells us that a deuterium nucleus is a proton plus a neutron. The same is true here. One of the major differences between the standard model and the non-singular potential is that within the standard model there is no real detail in the understanding of the constituents of the nucleus. The non-singular model gives equations of motion that describe the motion of the particles making up the deuterium nucleus. Once again I used the statement of conservation of spin and the magnetic moments to evaluate the effective units of actions for the protons orbit within the deuterium nucleus. The protons orbital radius within the deuterium nucleus may be found by comparing the predictions with the experimental mass.

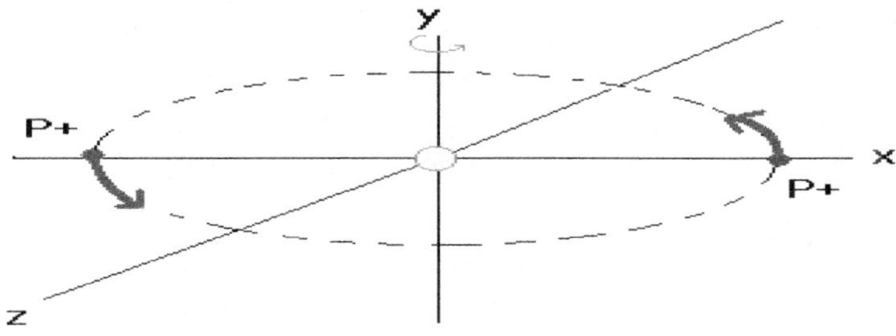

Figure 4.4 Two protons orbit around an electron.

Nuclear model

Next I investigated the possibility of stable orbits with three and four protons around a single negative charge. Three protons around a single electron would represent an isotope of helium called He^3. Four protons around a single electron would be an isotope of lithium known as Li^4. Neither of these isotopes is as common in nature as are He^4 or Li^7, but the equations lead naturally into a nuclear model. In order to see this nuclear model suppose we consider the fusion of two deuterium nuclei into a helium nucleus.

I would like to present a brief description of the crude nuclear model that seems to grow from this approach. Perhaps the best way of allowing this model to be visualized is to present a description of the fusion of two deuterium nuclei into a helium nucleus. Two deuterium nuclei may be caused to approach each other in various ways. It should be noted that the two protons are spinning around the electron each keeping directly opposite from the other because they hate each other. This means that if I put a pencil through the electron that is perpendicular to the plane of the protons' orbits the pencil will be in line with the orbital spin axis. Now if I slowly nudge these two deuterium nuclei toward each other being careful to keep their orbital spin axis aligned the two protons of one deuterium nucleus will align themselves as far as they can from the two protons of the other deuterium nucleus; again because they hate them. See Figure 4.5. In the figure you may notice that the two electrons tend to move out of the plane of the protons' orbits away from the other electron. Remember the electrons also don't like each other.

As the two deuterium nuclei continue to be pushed toward each other the orbital radii of the protons tend to get a little larger and the electrons tend to move a little farther from the plane of the protons' orbits. However,

114

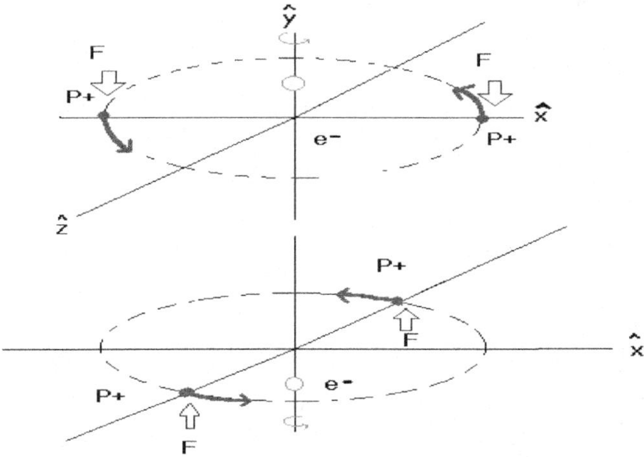

Figure 4.5 Two deuterons being pushed together.

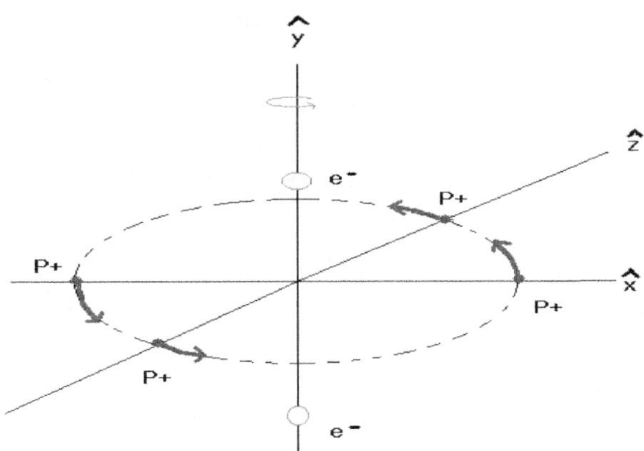

Figure 4.6 A helium nucleus.

the resulting helium nucleus, Figure 4.6, when the deuterium nuclei are pushed together so that the protons all share the same plane of rotation, is very stable because all four protons are attracted to the two electrons. It is true that the protons try to maintain their distance from each other, but their orbits are at a point where the repulsion between the protons is a little less than the coulomb repulsion would be at that separation. It is also true that the electrons don't like each other and at this separation they still have pretty much the Coulombic repulsion.

There are a few things to notice about the helium nucleus that will help visualize the nuclear model. First, notice that there is symmetry in the protons' orbits. Because of the love/hate relationship among the protons and the funny love you/hate you attitude of the electrons for the protons, symmetry must be maintained else the center of mass begins to move. Motion of the center of mass means the nucleus becomes unstable and will not long last. Next, notice that both the deuterium and the helium nuclei have the interesting feature that the charge is opposite that of atoms. In atoms the lighter electrons were on the outside, here the heavier protons are on the outside. This is due to the non-singular potential, but is totally opposite to our intuition and prior teachings we learned in school. The curious negative charge in the middle and positive charge on the outside is the aspect of the non-singular potential that allows a nuclear model to be developed that predicts the masses of the nuclei better than the current semi-empirical mass formula of the standard model.

The model is built up by taking the number of negative charges in the center of the nucleus to be the new number, Y. Z is the usual charge of the overall nucleus, and A is the familiar mass number. However, since the equations for the energies are transcendental equations, I used the symmetrical, 2-proton energy equation and just allowed the additional protons to go into a higher quantum numbered orbit just as the electrons fill the atomic shells. The accompanying table, Table 4.1, lists the Y, Z, and A numbers along with the experimental mass, predicted mass, the error between the predicted and experimental masses, the error on a per-mass number basis and the predicted binding energy per nucleon.

116

It may be noticed that since I simplified my approach by using the 2-proton energy equation those nuclei that have multiples of three protons have greater error in the predictions. This should be expected from the symmetrical approach. I accepted this error as it is less than the error of the semi-empirical mass formula and allows visualization of how the nuclear model builds the table of nuclei.

It must be stated that this model should be revisited with solutions of the modified Dirac equations (discussed below) in order to benefit from the full relativistic solutions.

Table 4.I. Experimental and predicted nuclear masses.

Y	Z	A	Exp. Mass (MeV)	Predicted Mass (MeV)	Predicted (MeV)	$\Delta M = E_p - E_E$)M/A (MeV)	BE/A (MeV)
1	1	2	1875.0	1873.7	-1.3	-0.7	2.1
1	1	2	1877.9	1879.7	1.8	0.9	1.2
1	2	3	2808.3	2819.9	2.6	0.9	1.7
1	3	4	3749.5	3748.2	-1.3	-0.3	1.5
2	1	3	2808.9	2804.2	-4.7	-1.2	4.4
2	2	4	3727.3	3736.1	8.8	2.2	4.9
2	3	5	4667.5	4668.1	-4.8	0.1	5.2
2	4	6	5604.7	5600.0	-4.7	-0.8	5.4
3	2	5	4667.8	4668.3	0.5	0.1	5.4
3	3	6	5601.4	5600.0	-1.0	-0.2	5.5
3	4	7	6531.8	6532.3	-1.6	0.1	5.6
4	1	5	4691.8	4689.2	-1.6	-0.3	1.5
4	2	6	5605.5	5610.4	4.9	0.8	4.1
4	3	7	6533.3	6531.6	-1.6	-0.2	5.9
4	4	8	7454.3	7452.7	-1.6	-0.2	7.3
5	2	7	6545.7	6545.7	0.6	0.1	4.6
5	3	8	7471.2	7471.2	-1.5	-0.2	5.9
5	4	9	8392.2	8395.5	0.8	0.1	--
6	2	8	7482.5	7482.5	1.0	0.1	3.8
6	3	9	8406.7	8404.7	-2.0	-0.2	5.3
6	4	10	9325.0	9326.0	1.0	0.1	6.5

Yang-Mills and SU3

The standard model does not have a functional form for the weak or the strong force. Many researchers have spent a lot of time and effort trying to learn how to describe the interaction of these forces. This research has resulted in obtaining equations that, if solved should describe the interactions. For the weak interactions the equations adopted are called the Yang-Mills equations while the equations of the SU3 group are thought to describe the strong interactions. I was interested to know if these sets of equations could be found residing within the Weyl Gauge Principle.

In developing the nuclear model presented above I had stuck to the relativistic approximation that had been suggested for the atomic model by Rutherford. The most accurate description of the nuclear interactions should come from the Dirac equation as it is the fully relativistic statement of QM. However, Dirac's development also included the assumption that Newton's third law was to hold as had Schrödinger's development of his wave equation. This required that I redevelop Dirac's equations while not requiring Newton's third law. For someone knowing Dirac's original development this procedure is straightforward, as they say, but it was extremely tedious for me. There was what seemed like an infinite number of places where an error such as a reversed sign could be made. I think I succeeded in making at least half the errors that were possible to make. In the end I had the modified Dirac equations for the two-body problem stated as motion of the center of mass and motion about the center of mass. It was not difficult to see how these equations collapsed back into Dirac's equation when Newton's third law held.

Now I had the relativistic QM equations for the un-like particle interactions that describes the nuclear forces typically referred to as the weak forces. The expression that I had derived did not even come close to looking like the

Yang Mills equations of the standard model. How was I to show that the standard model was a subset of what I had derived? This was another time when I sat down at my desk and asked for God's help. Again I suddenly had a notion that a certain definition of fields might be worth looking into. I applied that definition and soon had the Yang Mills equations before me. This proved that these equations, which were taken to be the currently accepted description of the weak forces, were indeed the results of a relativistic QM approach when Newton's third law was not enforced.

The SU3 group is a set of field equations that are suppose to house information about the interactions of the three quarks making up the proton in the standard model. I felt I needed to show that these equations were also a subset of the non-singular gauge potential. In order to describe the interactions among three particles I thought I really needed to write the equations of relativistic QM for three particles in the form of motion of the center of mass and motion about the center of mass. It took a while to do this and the resulting equations were not pretty.

I again asked God to help me if He were so inclined. Within ten minutes I found a definition that allowed me to write the equations for the three particle system in a form that allowed one to see the SU3 group equations. I did not, and have not, tried to solve either the Yang-Mills equations or the three-body SU3 equations. I have somewhat of a mental block that makes solving Dirac's equation difficult. Both of these sets of equations were much more difficult than Dirac's so I am leaving this job to someone else to tackle later. I intend to stay with the relativistic approximation as it seems to give good qualitative answers and reasonably good approximations. These should be good enough to get some analytical understanding of what is going on in the nuclei.

120

Protons

The standard model says the proton is made up of three quarks. Certainly, the scattering data does show three small things inside a proton. However, quarks may not be the only things that can be used to understand the makeup of a proton. The non-singular potential with a lambda that is different for the different particles may also offer a version of the constituents of a proton.

Above we showed how the deuterium nucleus could be described by two protons in orbit around a single electron. The resulting nucleus had an overall electric charge of one positive unit though the nucleus was made up of three particles, two with a positive charge and one with a negative charge. If the proton scattering data showed there were three particles making up the proton, could these three particles be two positrons and an electron? This would be three particles and the overall charge would be correct at one positive unit.

In order to rapidly investigate the possibility of a proton being made up of two positrons and an electron, we could use the same equation as used for the deuterium nucleus after we find the effective units of action for the positron orbits and the lambdas for both the electron and the positron. To be accurate the formal solution of the relativistic quantum description of the three body problem must be found. However, an estimate may be made if both the effective units of action and the lambda of the positron are both scaled using the ratio of the positron to proton masses. This scaling would produce $\hbar'_e = 0.0001362\hbar$ and $\lambda_{pos} = 1.77 x 10^{-18}$ meters.

Why did I use the masses to scale down the lambda of the proton to use for the positron and the electron? Because when I applied the non-singular gravitational potential to the planetary orbits that will be presented later, the correct value for the lambda of the Sun was given by

$$\lambda_{Sun} = \frac{GM}{c^2}.$$
(15)

Also, when you scale down the lambda for the electron using the mass ratio of the electron to the proton you get $\lambda_e = 1.77x10^{-18}$ meter and this compares favorably with the fact that the electron-electron scattering appears to still be nearly Coulombic down to 10^{-17} meter.

Using these values, plus the masses of the electron and positron in the deuterium three-body relativistic equation for the total energy of the proton and setting the total energy to the experimental mass of the proton yields a radius for the positrons within the proton of $r_{pos} \approx 5.728x10^{-20}$ meters. Therefore, this argues that the non-singular potential gives a prediction that the proton could consist of two positrons in very small, sub-nuclear orbit around an electron.

Virtual Particles

When Yukawa proposed that a nucleon may emit a particle with an appreciable rest mass that hovers near the nucleon only to be reabsorbed by the nucleon, he opened the scientific community's eyes to the realm of virtual particles. In this realm the masses of the virtual particles may be found by using the Klein Gordon equation and the potential of the nucleon. A problem exists within the standard model as a functional form does not exist for many nucleons. The non-singular potential provides the needed potential for all the nucleons discussed individually above so mass of virtual particles for these nucleons may be predicted. In the standard model not even the mass of the proton can be predicted.

Virtual particles are particles that may exist only for a short time so as not to violate the conservation of energy principle too long as established by the uncertainty principle. However, some of these virtual particles can leave a trail in a detector if they are charged. This means that these particles might not live long, but long enough to be detected.

When I used the values for the neutron given above in the Klein-Gordon equation I found the virtual particle that holds the center of mass of the neutron together to have a mass of

$$m_N \approx 3.28x10^3 \, GeV\big/c^2 . \qquad (16)$$

This is a super-heavy weight particle in terms of virtual particles.

Next I calculated the mass of the virtual particle associated with holding the deuterium nucleus together. Using the values for the deuterium nucleus in the Klein-Gordon equation produced a more reasonable particle mass of

$$m_D \approx 30.5 \, MeV\big/c^2 . \qquad (17)$$

Being on a roll I then used the values I obtained for the positron's orbit in a proton and found that these virtual particles' mass were

$$m_P \approx 7,254 MeV \, / \, c^2 . \qquad (18)$$

This virtual particle mass is also a heavy particle mass. Both this mass and that of the neutron's virtual particle mass are indicative of the small orbits of the positrons in the proton and the protons in the neutron.

Particle zoo

When I found that the universe was fundamentally a five dimensional manifold, as will be presented in the next chapter, I proceeded to develop a full five dimensional relativistic quantum mechanics. This development was filled with many complicated equations and new concepts. It was difficult to wade around in this world and make sense of it. I am certain there are many useful concepts and predictions that will come out of this world. However, for the moment I only wish to present two predictions that may prove relevant.

When, during the development of the five dimensional QM, I reached the spin vectors, I found that instead of just the three usual components of spin as one would have in three dimensional space, there were three spin vectors. First, there was the usual three component spin vector. Secondly, there was another, new three component spin vector and then there was a new four component spin vector. I have not come close to understanding these new spins, but one, the new three component spin vector seems to be tied to the magnetic moments of gravitating bodies.

What is relevant here is a theorem that can be proven concerning an interconnection among the ten components of spin. The results of the theorem are that the only combinations of the ten spin components that are allowed are certain octets. This seems a lot like the fact that particles in the particle zoo only come in octets.

Chapter 5 Gravitation

Previously the fundamental laws of thermo-dynamics were adopted as foundations of mechanics as well as thermodynamics. The previous chapter displayed how these laws provided both geometry and equations of motion. The Weyl Gauge Principle and Weyl Quantum Principle were presented and some of the results from these principles discussed. However, in these chapters the gravitational field received little attention. Now gravity will have its time.

When I was teaching classical thermodynamics at the Naval Academy it dawned on me that in writing the first law on the board not only did I write a statement of the conservation of energy, but I also wrote the foundations of a five dimensional universe. I could not find a way of proving the contention that mass density could always be written as a function of space and time. So I adopted the reverse position and sought to find some prediction that differed from experimental findings. I have now studied the Earth's gravitational field and its magnetic moment and the briefly reported, or claimed, fifth force of nature. I studied the planetary orbits and cosmology. Most recently I studied the reports of the hypothesized dark matter and dark energy. Nowhere in these studies have I found a place where experimental data differs from the predictions that come from assuming mass density is fundamentally independent of space and time.

However, by starting with TD laws and proceeding to develop the five dimensional gauge field and force laws, any presentation always seems to leave the audience waiting for me to compare this with Einstein's general theory of relativity. I always felt that Einstein's general theory was just a subset of the five dimensional gauge theory, but it never was so easy to see as it was after I asked for God's assistance in proving my contention. The resulting development was so mathematically direct and, to me, overwhelmingly logical that instead of taking the reader through the long development before showing that Einstein's general theory is a subset of the five dimensional gauge fields I shall start with this proof and then return to the development of the five dimensional gauge fields.

Einstein's General Theory of Relativity

In the two decades after Einstein published his general theory of gravity and it was verified by experimentally measuring the bending of light around the sun many researchers sought to find a way of unifying gravity with electromagnetic fields. A lot of books discuss these efforts better than I could possibly do so I will only mention that a few of these researchers, including Einstein himself, looked into five dimensional manifolds for some possible connection. All of these early five dimensional efforts started with the thought that the fifth dimension cannot be real so the researchers used this as an assumption and forced many terms in the resulting field equations to zero so there would be no possible connection to anything that might argue that the fifth dimension might have a connection to reality.

The TD approach argues that the fifth dimensional is real and is the mass density. This argues that the conservation of mass might, indeed must, be violated somewhere in the universe. Einstein never seemed to have such a thought. His general theory was developed while assuming that mass was conserved. This then sets up our approach to proving Einstein's general theory to be a subset of five dimensional gauge fields. I will return to the WGP and simply assume that perhaps there is a fifth dimension that is real, but that this fifth dimension, without making any statements about its character, must be conserved in the same fashion that we have always required mass to be conserved. But before we do this I would like to point out that the WGP of course talks only about gauge fields and these are fields that depend upon the Weyl scale factor varying in accordance to a gauge function. This is very different from Einstein's curved Riemannian manifold in which he used gravity to curve the manifold. I hope you watch to see how the conservation of the fifth dimension changes a "length curvature" controlled by WGP to a vector curvature controlled by mass.

It was this approach that I thought of shortly after I had asked for some assistance from God. Once I had this approach in mind the math, though very highly involved with equations that had many components, was straight forward and took less than half a page to express.

A four dimensional gauge field has six non-zero components which are the three vector components of the electric field and the three vector components of the magnetic field. A five dimensional gauge field has ten non-zero components. The same six components of the electromagnetic field, a new three component vector field plus a new scalar term. We should have expected a three component vector field because we know that is the character of the gravitational field. The new scalar term throws a different wrinkle into the fields that has no predecessor with which to compare it.

Not to be thrown off track by some new aspect of physics, we may simplify the problem a great deal by assuming that the electromagnetic components of the gauge field are zero so we only address electromagnetically neutral bodies. We should also be able to reduce the new vector field to only one non-zero component. For the moment lets also assume the unknown new scalar term is zero. Now we have only one non-zero term to think about. In the process of deriving Einstein's general theory this reduction of the gauge field is not necessary. However, it is necessary if you then wish to write out all the field terms in order to compare them to what we previously had.

Next we may use the restriction that the fifth dimension must be conserved to write the fifth dimension in terms of space and time with an equation that is similar in form to the statement for conservation of mass. This means that the fifth dimension can be neither created nor destroyed but may flow around in space and time and increase or decrease its density. This becomes a restriction on the five dimensional gauge field and this restriction embeds a four dimensional hyper surface into the five dimensional manifold. This is the same concept that we run into when we restrict ourselves to a surface that is everywhere the same distance from a specified point in three dimensional space. Of course you will recognize this restriction means we are talking about the surface of a sphere. But notice that the background three dimensional space need have no curvature while the sphere is curved. The restriction of equal distance from a point imposed the curvature on the sphere.

The same thing is true of the restriction of conservation of the fifth dimension. This restriction imposes a curvature onto, or into, the four dimensional hyper surface. The expression of this curvature is a equation just like Einstein's field equations of his general theory. Further, the curvature depends upon the fifth dimension! If we think that Einstein's general theory is close to the mark in describing the nature of gravity then we have no choice but to name the fifth dimension mass density. Therefore, Einstein's general theory results when we look at the five dimensional gauge fields given by the WGP and we impose conservation of mass!

This is what I had wished to show for years, but could never find the right starting point until I asked for divine assistance. You might claim that I had worked with these equations, thought about these concepts and had lived

128

with the desire all these years and that finally my mind put it all together. I could not put forth a strong convincing argument to the contrary. However, in my own mind, since four times I had asked for help and four times I quickly found what I sought, I have little doubt that I had indeed received assistance.

The Inductive Coupling

The five dimensional gauge field requires that there be a tie between the electromagnetic and the gravitational fields. This is the same as the inductive coupling that is required between the electric and the magnetic fields. This coupling between the electric and magnetic fields is the reason they are referred to as the electromagnetic field. The name implies a tie between them. Perhaps the most useful and well known result of this tie is our system of electrical power distribution. Direct current can be generated and used in motors, but alternating current can have its voltage and current changed by putting it through a transformer. Raising the voltage reduces the current and greatly reduces the losses of power transmission lines. This great reduction of losses is the primary reason our electrical power distribution system is based on alternating current rather than the direct current system Thomas Edison developed.

The strength of this coupling between the fields may be found when the various field components are described in identical units. The electric field is usually stated and measured in units of volts/meter. The magnetic induction has units of weber/meter^2 and the magnetic permeability constant, μ_o, is used to convert these units into units that match those of the electric field. When the magnetic inductance is multiplied by the permeability constant the result is called the magnetic field and is given in units of volts/meter. This then allows the gauge field to have all components written in terms of volts/meter.

In order to include the gravitational field in this gauge field there needs to be a unit conversion that will convert the usual units of the gravitational field to volts/meter. By studying the interrelationship of all the gauge field components it may be shown that the necessary conversion factor is $\beta \equiv \sqrt{\varepsilon_o G}$ in units of coulombs per kilogram. This was a curious combination of already measured quantities. But it was a welcome sight as I had feared that a new constant may be needed and not already measured.

Magnetic moment of the Earth

The charge-to-mass conversion factor, β, may be quickly put to a test. The inductive coupling between the gravitational and electromagnetic field means that a gravitating body that is electrically neutral should have a magnetic moment. This argues that if we take the spin of the Earth and multiply the mass of the Earth with the conversion factor we should be able to predict the magnetic moment of the Earth. This method of predicting the magnetic moment provides a prediction of

$$\mu_E = 8.6 \times 10^{22} \, amp\text{-}m^2. \tag{19}$$

This predicted value of the earth's magnetic moment compares very well with the experimental value of 8.1×10^{22} amp- m^2 especially when the prediction was made by assuming the mass density is uniform throughout the Earth which, of course, it isn't. However, to my way of thinking this is a much better explanation for why the Earth has a magnetic field than trying to explain it with currents circulating within the Earth and then needing to explain what causes the currents.

Gravitational Rotor

The prediction of an inductive coupling between the electromagnetic and the gravitational fields argues that we should be able to electromagnetically create a gravitational field. It also argues that we should be able to gravitationally make an electric field. Because I felt we could manipulate the electromagnetic field better than we can manipulate a gravitational field, I began to think about how we might electromagnetically create a gravitational field. Some have called it an anti-gravity experiment. I called it a gravitational rotor and it was designed to make half of a rotating element heavier and half of the element lighter and thereby causes the element to rotate in the Earth's gravitational field. Another line of investigation actually uses the opposite route and seeks to measure the electric field predicted to be the result of manipulation of a gravitational field. This line of experimentation has more, as yet unpublished, success than my gravitational rotor.

In looking through the new equations, I noticed that the new conservation of charge equation had one more term than the old one. The appearance of the new term suggested that if an electric current were forced to diverge as in flowing into the apex of a thin cone and out the base, which had a larger diameter, then the new term would have to be non-zero. The new term seemed to be a gravitational current density.

I didn't know what a gravitational current density should be, but I did have a person working for me that could make almost anything. So I asked Tom Gould if he would make me a set of cones out of thin copper attached at their bases and put on a pivot so they could rotate. The cones were lying on their sides with their apexes pointed to each side while being allowed to rotate so the apexes moved up and down. In each apex he put a brass rod that was bent so that about an inch and a half of each rod pointed downward. These vertical ends of the brass rods where placed in baths of mercury so current could be made to flow into one mercury bath, into the first brass rod, into the first cone, diverge within the thin walls of the cone to the base of the first cone, pass into the second cone's base, converge as it flowed through the second cone to it's apex, into the second brass rod, into the second mercury bath, and back to the battery. In this fashion the first cone caused a divergent electric current which created a positive gravitational field that opposed the Earth's gravitational field while the second cone forced a converging electric current that created a negative gravitational field that attracted the Earth's gravitational field. Thus, one cone was made heavier and the second cone made lighter causing them to rotate. Their rotation caused one brass rod to be driven down into its mercury bath and the other rod to be pulled out. The difference in the displacement of the rods in the mercury gave a calculation of the value of the torque produced in the cones by the flow of the electrical current.

We tested this means of electrically creating a gravitational field one evening in our home where we could have a large room with 30 foot distance between the cones and a distant wall. On the rotating shaft Tom had glued a small mirror so we could shine a small laser onto the mirror and have the spot appear on the distant wall. By the spot we placed a vertical ruler so if the cones rotated the ruler would show its motion by displaying how far the spot moved.

When we turned on the current the spot moved about 4.5 inches as recorded by someone putting their finger on the new location. When we reversed the current flow the spot moved in the opposite direction. The amount of current we used to get this approximately 0.030 radian movement was 600 amperes and this caused a slight bubbling in one of the mercury baths.

The argument might be made that this much current flowing from a liquid conductor to a solid conductor might cause a torque without it being produced by the expected creation of a gravitational field. Two lines of activity followed. First, I passed the gravitational rotor Tom Gould had made to folks at the Los Alamos National Laboratory so they could do an independent test of the rotor. Secondly, Timothy Rynne and John Dering of Science Applications and Research Associates, Inc. (SARA) of Huntington Beach, CA built a device to replicate the experiment.

The gravitational rotor that was sent to Los Alamos National Laboratory became lost without any explanation. The device constructed by SARA was tested and the results reported in The Electric Spacecraft Journal. The results I had obtained seemed dubious to me in light of possible interaction between the liquid and solid conductors. SARA's device sought to remove this possibility by putting their power source on the rotating element. The results SARA obtained seemed to involve the appearance of an armature reaction similar to that experienced on all electric motors. Both results did not seem to warrant additional time away from making a living and were dropped at that time. I have now gotten a paper accepted for presentation at the upcoming Space, Propulsion and Energy Sciences International Forum that includes the prediction of the gravitational rotor in the form of space craft propulsion.

Time Dependent Gravity and Fundamental Constants

There are now several fundamental constants used in physics. Some of these are the speed of light, c, Planck's constant, h, the unit of electric charge, e, the gravitational constant, G, and the time dependence of the gravitational field, $\dot{G} \equiv b$ which is a new constant that shows up in the five dimensional gauge fields. It has been asked in the literature if all these constants are really fundamental. This is especially true if one includes the long list of constants needed in the standard model of particle physics. The five dimensionality of the gauge field offers an additional point of view with respect to the independence, or fundamental character of some of these constants.

While I was going through the five dimensional gauge field tensor looking for a statement about the unit conversion needed for the gravitational field, I discovered a different look at a few of these fundamental constants. This group of constants includes the newly discovered time dependence of the gravitational field, $\dot{G} \equiv b$, required by the five dimensionality of the field.

This section contains some very interesting points (at least I think they are interesting) that the reader may like to see developed in a little more detail than I have been doing. Primarily I have tried to stay away from equations as typically they are complex and perhaps difficult to understand if not familiar with them. The equations in this section are not quite so complicated and are perhaps a little easier to take.

This discussion is intended to display the manner in which the Dynamic Theory addresses the atomic and nuclear states and provides a basis for interpreting the more difficult full investigation of these states. The basis rests upon the WQP's quantization of a gauge function, which yields quantization of electric charge and orbital like interaction of particles, and the premise that the functional form of the gauge function remains the same in the atomic and nuclear realm as it is in the cosmological, or red shift, realm. What I would like to give you in this section is a small appreciation for how important of a role the gauge function plays in the whole of physics.

As we go through the process of learning about electromagnetism, we hear a lot about there being a gauge function, but it is more the aspect that the gauge function of electromagnetism is not completely known. This means there are some things about the gauge function we don't know, and even that we don't need to know. The important thing about the gauge function of electromagnetism that we needed to know was that electromagnetism must be gauge invariant. By this it was meant that there were ways in which one could transform the problem from one coordinate system to another in which the physics of the problem remained the same. Gauge invariance implied that in any transformation the gauge must remain the same. This was thought to be so important that one of the first questions asked of any new hypothesized theory was, "Is it gauge invariant?" Here I wish to let you see that we can, and must, go beyond knowing if the theory is gauge invariant, but we need to know about the gauge function itself. This is sort of a no-no in classical electromagnetism as we were told that the real physics was in the electric field. It was the electric field that showed up in the force and it was the force that we could measure. So the only reason to discuss the electrostatic potential was that it shaped the force. We could never measure the potential so there was no real reason to learn about the function that shaped the potential. I'm arguing here that there is real reason to learn about the gauge function that shapes the potential that, in turn, shapes the force.

Time dependence

Our starting point is the gauge function in the form determined by a study of red shifts, this form is

$$\ln f^{\frac{1}{2}} = mK\left(1-bt\right)\frac{e^{-\frac{\lambda}{r}}}{r}. \tag{20}$$

In the gauge function of Equation (20) please note that it depends upon space, time and mass. The reason it involves space, time and mass is that it is a gauge function in a manifold of space-time-mass and, therefore, must have all three involved in its gauge function. I hope to explain a little bit about why this must be so.

The route to be taken toward evaluating the parameters of this gauge function includes looking at isentropic states in order to see how stable states reveal aspects of the gauge function. First, consider the WQP, which I have not displayed in its mathematical glory yet, but it has been discussed and argued to be the root of most, if not all, the atomic and nuclear physics. The primary reason it dominated the world of the small and super small was that it led to things we could look at, which were, by and large, the things that stayed around for a while like electrons and protons. I told you the WQP quantized the gauge potentials and this is how it does that.

First, you may have noted that in Equation (20) above I told you that the gauge function had a certain form and then I presented an equation that had defined the square root of the logarithm of the gauge function instead of the gauge function itself. I did this for three reasons. The first reason is out of habit, the second reason is that to write the gauge function by itself is another complication that is probably not needed, and the third reason is that the gauge potentials are defined in terms of how the left hand side varies with respect to the different variables. The way that the WQP quantizes the gauge potentials is that it forces the sum of the values of all the gauge potentials along a given path to be quantized. This requirement may be stated as

$$\int \phi_i dx^i = \int \phi_0 dx^0 + \int \phi_1 dx^1 + \int \phi_2 dx^2 + \int \phi_3 dx^3 + \int \phi_4 dx^4 = 2\pi N . \quad (21)$$

In Equation (21) I first wrote the line integral of the gauge potentials in a condensed form and then I expanded it so you could see each of the terms and see that we must consider how the potentials behave with respect to all five variables in the space-time-mass manifold. The WQP states that when we do this the answer must be quantized in integer values as specified by the letter N on the right hand side.

136

Now we want to know what gauge potentials are allowed for particles (stable states) and then what paths are allowed for interacting particles.

Taking the functional form of Equation (20) and looking at how it changes with respect to each of the space-time-mass variables, the only non-zero gauge potentials, assuming spherical symmetry, become

$$\phi_0 = \frac{\partial \ln f^{\frac{1}{2}}}{\partial (ct)} = -\frac{N_0 kbm}{c}\frac{e^{-\frac{\lambda}{r}}}{r}$$

$$\phi_r = \frac{\partial \ln f^{\frac{1}{2}}}{\partial (r)} = -N_r km\,(1-bt)\left(1-\frac{\lambda}{r}\right)\frac{e^{-\frac{\lambda}{r}}}{r^2} \qquad (22)$$

$$\phi_4 = \frac{\partial \ln f^{\frac{1}{2}}}{\partial \left(\dfrac{Gm}{c^2}\right)} = N_4 k\,(1-bt)\frac{e^{-\frac{\lambda}{r}}}{r}\;.$$

where the N's are the gauge potential quantum numbers required by the quantum condition and are contained within the K.

Don't worry about the math procedures by which I got these relations. Rather look at what may be different in the right hand side of these potentials as compared to the right hand side of Equation (20). In the electrostatic potential, ϕ_0, there is no time dependence. It got lost when we looked at how the electrostatic potential varied with respect to time; it doesn't. Next the vector potential, ϕ_r, reflects a more complicated looking dependence on r than appeared in Equation (20), but the dependence on time and mass remain the same. While you may notice that in the remaining potential, ϕ_4, there is no dependence on the mass, m, I really won't use this potential in the remainder of this section since it tells how mass must change and I want to stay with conservation of mass for the moment.

The gauge fields may be obtained by looking at the difference in two variations and checked within the eight field equations. We find that the non-zero field quantities are

$$E_r = \frac{\partial \phi_r}{\partial(ct)} - \frac{\partial \phi_0}{\partial(r)} = \frac{Z\eta kbm^2}{c}\left(1 - \frac{\lambda}{r}\right)\frac{e^{-\frac{\lambda}{r}}}{r^2}$$

$$V_r = \frac{\partial \phi_r}{\partial(\frac{Gm}{c^2})} - \frac{\partial \phi_4}{\partial(r)} = \frac{Z\beta\eta km}{\frac{G}{c^2}}(1-bt)\left(1 - \frac{\lambda}{r}\right)\frac{e^{-\frac{\lambda}{r}}}{r^2} \qquad (23)$$

where the η and β are included in order to obtain the correct units and Z is the difference in the gauge potential quantum numbers. β is the charge to mass ratio discussed in the previous section and η is yet to be determined. The appearance of the additional mass term in the fields is due to the fact that the field quantities are written in terms of field densities.

When the gauge fields are put into the field equations one finds that the electric charge is given by

$$q \equiv Ze = \frac{Z\varepsilon_0\eta kbm^2}{c}. \qquad (24)$$

While the gravitating mass is found to be

$$M = \frac{\varepsilon_0 Z\eta km}{\beta\left(\frac{G}{c^2}\right)}(1-bt). \qquad (25)$$

From Equations (24) and (25) one may notice that the electric charge is not a simple primitive quantity and that the gravitating mass is time dependent. The fact that the electric charge may be written in terms of other parameters indicates that there must be some connection between these parameters and the electric charge that we did not previously know about. The time dependence of the gravitational mass we will see later several times. The requirement that gravity must be time dependent is one of the important points to remember about this section.

It is also worth remembering that there could be no electric charge if gravity were not time dependent. See that in Equation (20) the coefficient in front of the time is the parameter I called b and earlier I gave the definition that $\dot{G} \equiv b$. If gravity does not depend on time $b=0$ and then by Equation (24) there could be no electric charge.

Fundamental constants

Comparison of the above fields and charges with the forces of interaction is a way in which the above equations may be checked against experiment. In particular, the atomic energy states are an attractive means of comparing with experiment since the atomic energy states determine the experimental frequencies seen from stimulated atoms.

From the force law which multiplies the charge of one particle, say particle 1, times the electric field of the second particles, say particle 2, we find that the electric force between the particles is given as

$$F_{e12} = q_1 E_{r2} = \frac{Z_1 Z_2 e_1 e_2}{4\pi\varepsilon_0}\left(1 - \frac{\lambda_2}{r}\right)\frac{e^{-\frac{\lambda_2}{r}}}{r^2} \qquad (26)$$

where we have not imposed the requirement that unit of electric charge be the same for all particles yet. Note also that should the separation, r, be significantly greater than the λ of the particle the radial dependence approximates that of classical electric forces.

The Dynamic Theory gives the gravitational force in the same manner as the above electric force between charged particles. Namely this means we must multiply the mass of body one times the field of body two and the results may be written as

$$F_{g12} = M_1 V_{r2} = Z_1 Z_2 G M_1 M_2 \left(1 - \frac{\lambda}{r}\right) \frac{e^{-\frac{\lambda}{r}}}{r^2} \qquad (27)$$

when Equation (25) is used and it is remembered that $\beta \equiv \sqrt{\varepsilon_o G}$ is the charge to mass ratio. This also looks to have a form comparable with the classical case. That is, when we look at large distances compared to lambda it collapses to the form of the Newtonian gravitational force.

To investigate the isentropic interaction of two particles the Dynamic Theory tells us to look at isolated systems where the entropy is constant. In this case the quantum condition must hold even if knowing the particles involved provides us with knowledge of the gauge potentials. What remains then is to ask what paths are possible given the gauge potentials are known and the quantum condition must be satisfied. While recognizing that paths are possible where dr is non-zero, suppose one considers circular motion for which $dr=0$. This is an allowed path and represents those chosen by Bohr in his first cut at atomic theory.

The full five dimensionality of the Dynamic Theory requires a comment on the fourth component of the gauge potential. For the moment we shall consider that $dm=0$ in our quantum condition. This is the statement that mass must be conserved.

From the quantum condition and the circular orbit assumption we find that

$$\int \phi_0 d(ct) = 2\pi N = -N_0 (kbm)_p \frac{e^{-\frac{\lambda}{r}}}{r} cT . \qquad (28)$$

In Equation (28) we note that the N is the orbital quantum number while the gauge potential quantum number, N_0, also appears explicitly. Also, it should be noted that the subscript, p, denotes the fact that the coefficient of time in the gauge function may be specific to the particle.

140

Now one may use the equilibration between the force of attraction between the particles and the centripetal acceleration to arrive at an expression for the velocity in the circular orbit of

$$v = \sqrt{\frac{-Z_e Z_p e_e e_p}{4\pi\varepsilon_o m_e r}} \qquad (29)$$

where the unit of charge has not yet been equated and the separation is taken to be large with respect to lambda. I have also used the subscripts e and p to indicate an electron and a proton. Using the relation between the period, radius and velocity of circular motion, we may find that the radius is given by

$$r = \frac{N^2(-Z_e Z_p e_e e_p)}{N_o{}^2(kbm)_p{}^2\,4\pi\varepsilon_o m_e}\,. \qquad (30)$$

We may now use the definition of angular momentum and Equations (29) and (30) to arrive at orbital angular momentum

$$L = m_e v r = \sqrt{\frac{-Z_e Z_p e_e e_p m_e r}{4\pi\varepsilon_o}} = \frac{N}{N_o}\hbar \qquad (31)$$

where the unit of action, or Planck's constant, is

$$\hbar \equiv \frac{-Z_e Z_p e_e e_p}{4\pi\varepsilon_o(kbm)_p}\,. \qquad (32)$$

We find that the unit of action in the interaction between the electron and proton is not a primitive number itself. Rather it depends upon the unit of charge and the coefficient of time in the gauge function of the proton. However, should one take into account the definitions of the particles' charges as given by Equation (24), they find that the unit of action depends upon the mass of both particles and the coefficient of time of the gauge function of the electron. To see this we should attempt to evaluate these coefficients.

We may use the potential and kinetic energy expressions in the same manner as is done in classical physics to arrive at the expression for the energy states. These will also be the entropy states since the energy and entropy differ by the relativistic factor and for these low velocities this becomes unity. The entropy states are thus given by

$$E = \frac{-m_e Z_e^2 Z_p^2 e_e^2 e_p^2}{2(4\pi\varepsilon_o)^2 \hbar^2 \left(\dfrac{N^2}{N_0^2}\right)}$$

$$= \frac{-m_e N_0^2 (kbm)_p^2}{2N^2} \tag{33}$$

where the last expression is obtained by using the definition for the unit of action, Equation (31).

Note that the energy states are determined by the mass of the orbiting particle, the coefficient of time in the gauge function of the field particle, the quantum number of the orbit and the quantum number of the electrostatic gauge potential of the field particle. The experimental evidence of these energy states, as given by the frequencies of excited atoms, will determine the value of the coefficient of time in the proton gauge function. All other parameters, such as the unit of electric charge and the unit of angular momentum may then be determined from this value.

The picture of the energy states may be further illuminated when one uses the determination of k from the red shift data. From red shift data the Dynamic Theory requires that

$$k = -\frac{G}{c^2} \tag{34}$$

so the expression for the energy states may be written as

$$E = \frac{-m_e N_0^2 G^2 m_p^2 b_p^2}{2c^4 N^2}. \tag{35}$$

From this it may be seen that the only parameter to be evaluated from experimental frequencies is the *b* itself.

Looking back at Equation (26) one sees that the force between the particles goes to zero when the separation equals lambda and then changes sign as the separation decreases further. This sets up additional possible circular orbits in which protons may orbit around electrons. This is the neutron we discussed before. We have developed the tools to quickly look at this case as well.

If one supposes that a proton is in orbit around an electron at separation of the lambda of the proton the separation is still much greater than the lambda of the electron and therefore, our previous assumptions which allowed the dropping of the exponential term holds in this case also. However, in using this case and the simplicity of the previous development of the atomic circular orbits we are ignoring the fact that in these tight orbits of protons around electrons the velocities are becoming relativistic. Since we wish to use the example as illustrative instead of exact quantitatively, we shall accept the errors the approximation introduces.

The proton is now the particle in motion so that its velocity may be given by Equation (29) replacing the mass of the electron with the mass of the proton. Similarly, Equation (30) specifies the radius of the proton when the roles of the electron and proton are reversed. The same is true of the expression for the angular momentum, so that the unit of action for this case must be given by

$$\hbar_e \equiv \frac{-Z_e Z_p e_e e_p}{4\pi\varepsilon_o (kbm)_e} \tag{36}$$

which has the same form as before, but different value.

In this case, calculation of the potential entropy/energy may be seen to be

$$V = -\int_r^\infty \frac{Z_e Z_p e_e e_p}{4\pi\varepsilon_o r^2} dr \tag{37}$$

143

which means that when one adds the kinetic energy to the potential energy of Equation (37) one obtains

$$E_n = \frac{3m_p N_0^2 (kbm)_e}{2N^2} \qquad (38)$$

which may easily be seen to be positive. The positive value of the entropy/energy levels means that this state will decay through the mechanism of quantum tunneling. Previously, this state has been seen to be best equated with the experimental data of the neutron. The arguments that might be envisioned with regard to the unit of action may be answered by considering the difference between Equations (32) and (36).

Previously we did not follow up on the consequences of requiring the unit of electric charge to be the same for each particle, however, if we now impose this requirement Equations(24), (32) and (36) require that

$$e_e \equiv e_p \Rightarrow \hbar_e = \left(\frac{m_e}{m_p}\right)\hbar_p \qquad (39)$$

which is approximately the value determined previously in the neutron orbital effective units of action.

Suppose we now go to another example still accepting the approximations of non-relativistic motion and circular orbits. For this case consider two protons in binary orbits where the reduced mass would be half the proton mass. Since the field particle carries the field of the proton the unit of angular momentum would be given by Equation (31) while the energy states would be given by

$$E_{He} = \frac{-m_p \dfrac{N_0^2}{2}(kbm)_p^2}{2N^2}$$

$$= \frac{1}{2}\left(\frac{m_p}{m_e}\right)E_a \qquad\qquad (40)$$

$$\approx -12.5KeV\left(\frac{N_0^2}{N^2}\right)$$

which would suggest that any photons emitted as the result of transitions between these states would result in gamma emission.

Conclusions

1. Within the above one sees that there is only one primitive parameter in the entropy/energy levels of the atom. Should one choose the coefficient of time in the gauge function as the primitive one then both the unit of electric charge and the unit of angular momentum may be determined. Alternatively, when considering the example of the proton-proton interactions the requirement that the unit of electric charge be the same for both the electron and proton provides the right level of gamma emissions for transitions between these energy states. Therefore, it may be desirable to consider the unit of electric charge as the primitive number and the coefficient of time in the gauge function and the unit of action as the derived quantities.

2. It may be noted that there is no requirement for the unit of electric charge to be the same for the electron and proton in order to predict the atomic energy/entropy states. This is a rather curious point and probably should be studied at some later time.

3. The unit of action for atomic states is different from the unit of action for the proton orbit in the neutron.

145

4. The lifetime of the neutron may be calculated using the relativistic solution to the proton orbiting around an electron to obtain the potential function needed for the tunneling calculations.

5. The proton-proton interactions display the right level of gamma emission for transitions between the excited nuclear entropy states.

I hope this section gives the reader an appreciation of why gravity must depend upon time and that all of the fundamental constants we now use are not fundamental but may be expressed in terms of other constants. This section also displays how the small and super small are tied to the large and cosmologically large realms of physics. Further, I hope that this section points out that just because we can't directly measure either the gauge potentials or the gauge function that doesn't mean we can't know a lot about them or how this knowledge of them can help us understand this universe in which we live. I suppose you might say that this section displays my repulsion towards the attitude that if one aspect of a theory might not be able to answer a particular question it should not be interpreted that we should not ask or cannot know the answer.

General Relativity

Einstein sought to find a way of expressing motion due to a gravitational field as motion in a curved space without a force instead of motion in a space, chosen at will, under the influence of forces. Previously, we saw that the curved space in Einstein's general theory was a four dimensional hyper-surface embedded into a five dimensional manifold of space, time and mass density. This means there are two equal ways of describing the effects of Einstein's gravity. One is by using forces in the five dimensional manifold while the other is using Einstein's field equations. It doesn't take much reading about

146

Einstein's general theory to get the idea that this is a fairly difficult route to take. What I would like to do here is to point out that the other route, which uses forces, might sometimes be easier, plus the gauge potential specifies the force to be used without the need to solve Einstein's field equations.

Perihelion Advance

One of Einstein's predictions from his theory of gravity was the advance of the perihelion of the planetary orbits. Einstein first used the special theory of relativity, but only half the measured advance was predicted. Therefore, the full general theory was needed to get the prediction right. What we want to do here is to look at what prediction of perihelion advance might come from the non-singular gravitational potential.

There exists a rather sneaky way of checking the prediction a new potential form might make with respect to the perihelion advance that can be used. It apparently was used a lot back just after Einstein used the general theory to make his prediction. Various researchers tried to guess at a functional form that might provide the same prediction as did the general theory and this was a way they come up with that would allow for an easy means of comparison.

The trick is to look at small oscillations about circular motion. This was easier than obtaining the full solution with very complicated equations. It required adding a term that contained information about the orbital angular momentum to the new trial potential form to get what was called an "effective" potential. Next you look at how this effective potential varies with respect to the radius twice. By looking at the variation of the effective potential with respect to the radius once, the first derivative, the result is zero because that is the circular motion. The second derivative gives the predicted oscillations about the circular motion. This result can now be compared to the results of the predictions of perihelion advance from the general theory. The prediction is not identical in the sense that the non-singular potential has the exponential term. However, for radial distances that are large with respect to lambda the first order term is dominate and the higher order terms are vanishingly small in comparison.

Comparing the first order term to Einstein's prediction shows them to be identical when

$$\lambda = \frac{GM}{c^2}. \tag{41}$$

Therefore, for lambda given by Equation (41) the prediction is identical to that given by Einstein's general theory so the advance of the perihelion cannot distinguish between the two theories.

Redshifts

Previously the dependency of the unit of action on the gauge function was discussed. The gravitational field is a gauge field in the five dimensional manifold. The unit of action, together with the frequency of light determines the energy of the photon that comes from a distant star. If the unit of action depends upon the gravitational gauge field and the gravitational field depends upon both the time and the gravitational mass, how would the frequency of the light coming from a distant star compare to light generated here on Earth?

Hubble found a linear relationship between the distance to a star and the shift of its frequency toward the lower energy, or red, end of the spectrum. All light from distant stars exhibit this shift in frequency. Einstein's general theory adds an additional shift in the frequency due to the gravitational field strength. This part of the frequency shift is due to the weight of the photon that leads to light bending around a gravitational body. There are some distant stars that pose problems for linear distance relation

148

with frequency. One type of star that seems out of sorts with this interpretation is called quasar, or queer star, as its red shift is very large meaning it is far away and yet it shines so brightly that there is no known mechanism of energy production that can explain its brightness.

It becomes interesting to learn how the time dependent, non-singular gravitational potential might predict any red shifting of light from distant stars. As stated above the unit of action and the frequency give the energy of the photon. But the unit of action depends upon the time and the gravitational mass. One may now ask, "If the energy of the light does not change during its travels from the distant star to the point of its reception, what would be the frequency shift of the light?" This question has an answer that includes looking at the difference in the unit of action at the point of generation and the unit of action at the point of reception. The gauge function was found to vary as given previously in Equation (20) which may be restated here for convenience

$$ ln \, f^{\frac{1}{2}} = mK \left(1 - bt \right) \frac{e^{-\frac{\lambda}{r}}}{r}. \qquad (42) $$

There are three things that can be different in the gauge function between the point of emission and the point of reception; the gravitational mass, the time and the distance from the gravitational mass. The gravitational mass and the distance from the gravitational mass is basically the gravitational field strength so there are really only two independent things that vary between the points of emission and reception. Also, since the speed of light is a constant there is then a direct relation between the time and distance.

The process of predicting the frequency shift is then to calculate the unit of action at the distant star and the unit of action at the point of reception. The reader may notice that I keep saying "point of reception" rather than the "Earth." I do this because the prediction of frequency shift shows a difference in the shift if the point of reception is different from the earth's surface. The Hubble Telescope also takes data on the frequency shifts of light from distant stars and the prediction shows that the Hubble Telescope should get a slightly different answer than would be obtained at the Earth's surface.

The equation that contains the predicted frequency shift was found to be

$$z = \frac{\Delta\lambda}{\lambda_e} = \exp\left\{\left(\frac{-G}{c^2}\right)\left[\frac{M_r e^{\frac{-\lambda_r}{R_r}}}{R_r} - \frac{M_e e^{\frac{-\lambda_e}{R_e}}}{R_e}\right] + \left(\frac{HL}{c}\right)\frac{\left(\frac{M_r}{R_r}\right)}{\left(\frac{M}{R}\right)}\right\} - 1 \quad (43)$$

where the subscript e stands for emission, the subscript r means reception while no subscript means the earth's surface. The time from emission and reception is given in terms of the distance the light traveled and the speed of light. Therefore, L is the distance to the star and H stands for Hubble's constant which is the coefficient of the time dependence in the gauge function.

The first term in the exponent of this expression is the prediction of frequency shift due to the gravitational field strength difference between the point of emission and the point of reception. If the exponential terms within this gravitational part are reduced to unity this expression reduces to Einstein's gravitational red shift prediction. The presence of these terms provides an answer to the question concerning the energy production of quasars. Recall that the lambda of the gravitating body was found to be proportional to the mass of the body. This means that the exponential term becomes important the more massive the body becomes. For distant bodies the more massive bodies will display larger red shifts. This means that for the more massive bodies the linear relationship assumed by Hubble and currently used by scientists does not apply. This means the quasars are not as far away as the Hubble relation would have them be if they are more massive than other stars. Further, the term is an exponential term and this means that there becomes a point in the mass of the body where any increase in mass makes a large increase in the red shift.

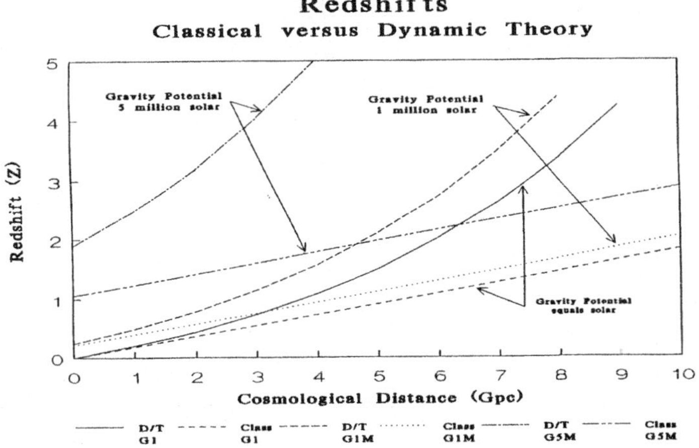

Figure 5.1 Comparison of red shifts.

The second term in the exponent involves the distance to the star. But it has a curious multiplicative term that is the ratio of the gravitational field strength of the point of reception to the gravitational field strength at the Earth's surface. The appearance of this ratio in the expression for the red shift due to the distance of the star was surprising to me and I thought it was an error. Dan Ross checked, rechecked, and doubled checked my result as I had, and found that this term should indeed be present. There should be a difference showing up here in addition to the difference of the predicted red shift in the gravitational part.

The presence of this term means that if we measured the red shift at the Earth's surface and at the height of the Hubble Telescope there should be a

151

measurable difference in the result. The height of the Hubble Telescope puts this difference in the 6% range and should be discernible. Once we had solidified this prediction, Dan Ross stopped by the headquarters of the Hubble Telescope on his next trip east and talked with a few folks there to see if they had seen such a difference. Dan and I, were disappointed to learn that they calibrate the difference away by calibrating all red shifts to the Earth's surface. This is a shame because it throws information away. But we could not get anyone to listen to our suggestions of considering separating the results.

The information that gets lost in not comparing the red shifts measured at the height of the Hubble Telescope may be seen from the following. There are two unknowns about the distant star appearing in the red shift prediction of Equation (43), the distance to the star and its gravitational field strength. A separate measurement of the red shift at the Earth's surface and at the height of the Hubble Telescope then gives us two equations in these two unknowns. By comparing the two red shifts we could solve this system of equations for not only the distance to the star, but also its gravitational field strength. Such knowledge could be used to check any unusually large red shifts of objects that also may have a possible physical connection to stars with lower red shifts. It is a shame that the strength of belief in the current theories is so strong as to rule out such a test of a potentially useful prediction.

Quantum gravity

Previously we presented the case for isentropic orbits which are solutions of Schrödinger's wave equation for un-like particles. If we combine the knowledge of the lambda for the sun and the Earth given by the relation in Equation (41) which is

$$\lambda = \frac{GM}{c^2},\qquad(44)$$

then the Earth's orbit around the Sun would be more accurately predicted by Schrödinger's wave equations for two unlike particles. This quantizes the planetary orbits. Because of the two body system there are some quantum numbers that are not available for planets to use. However, the quantization is both possible and consistent with the solar system data.

These equations provide the elliptical solutions familiar to planet orbits. There the semi-major axis is given by $a = \left|\frac{-GMm}{2E}\right|$ and the eccentricity may be found as $\xi = \sqrt{1 + \frac{2E_n^2\hbar_g^2}{m(GMm)^2}}$. For the quantum numbers n, l, and m_l, there are many possible quantum states to find the planetary orbits to be in. However, not all integers are allowed as quantum numbers. The solar system with all the planets and comets may also be considered as a multi-planet system with all the similar potential relations between the planets as has been seen for the multi-electron atoms. Here there is an increased complexity and richness of possibilities in the differing properties of each planet. These solutions may allow classifying stars in the universe as the atoms are now classified; that is, there may be a periodic table of stars.

Those who have not read of the data supporting the quantization of the solar system may wish to look in Dr. Halton Arp's book "Seeing Red" published by Apeiron (1998). The quantized orbits of the planets, taken individually as a single planet system, would show quantized semi-major axis as $a_n = \left|\frac{GMm\hbar_g^2 n^2}{\mu(GMm)^2}\right| = \left|\frac{\hbar_g^2 n^2}{\mu GMm}\right|$. If the experimental value of a_n for the Earth is taken with the Earth's quantum number, 5, then the value of the gravitational unit of action for the solar system would be

$$\hbar_g = \sqrt{\frac{a_5 \mu_E GMm_E}{25}} \approx 5.32 \times 10^{39} J - \sec .$$

Chapter 6 Cosmology

The time dependent, non-singular gravitational potential gives a somewhat different picture of the universe than the standard big bang model. One might guess that a gravitational field that has no singularity might also avoid a singularity in its picture of the universe. This guess would be right. There are two equal ways of describing the universe. One by Einstein's curved space and the other by the forces in the five dimensional manifold. Remember these are the same thing when mass is conserved. In any study of the universe that is five dimensional with mass being the fifth dimension, restricting one's attention to only those events for which mass is conserved might be considered unrealistically restrictive. However, this is the realm in which comparisons with the cosmology of Einstein's theory perhaps should be made.

The Expanding Universe

The non-singular gravitational potential is time dependent and gravity is getting weaker with time. All gravitational orbits should then be expanding. An extension of this might lead to an expanding universe. But a geometry that has entropy as its arc length then argues that, if the universe is all that is and there is nothing with which the universe can exchange energy, the entropy, or arc length, of the universe must be seeking a maximum. Since the entropy is the arc length, or distance between two points, and it must be increasing then the universe must be expanding and can never contract.

The non-singular gravitational potential may be used in a study of the universe's behavior just as the Newtonian gravitational potential is now used in presenting the big bang model of the universe. This is where the non-singular character of the gravitational potential gives quantitative predictions to go along with the qualitative understanding presented above. Here is where one sees that the non-singular potential leads backwards in time to a point when the expansion velocity was zero. At this point in time the universe was much smaller, but still had a finite size. Also, at this point there would have been a tremendous acceleration applied to the universe that would cause an extremely rapid expansion initially. Over time this expansion acceleration decreases to zero and then the universe expansion begins to decelerate.

This evolution of the universe may be plotted in terms of the size, expansion velocity and the deceleration of the universe.

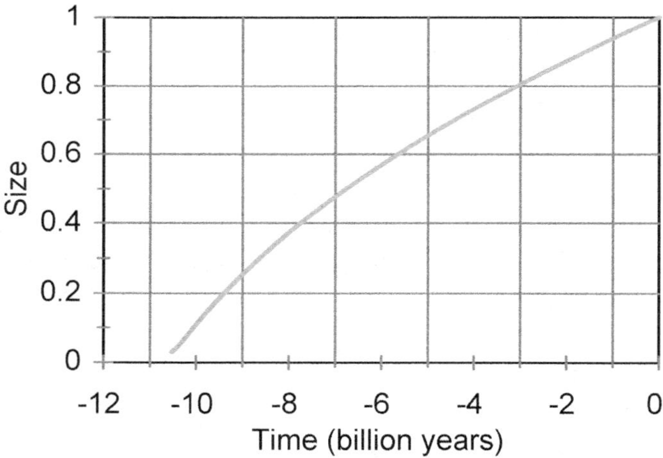

Figure 6.1 Cosmological Evolution

In Figure 6.1 the cosmological evolution is plotted starting at a finite, but much smaller, size some 10 billion years ago. The primary reason for this time to be different from current predictions of the big bang is the exponential term in the gravitational potential.

156

The expansion velocity is plotted in Figure 6.2.

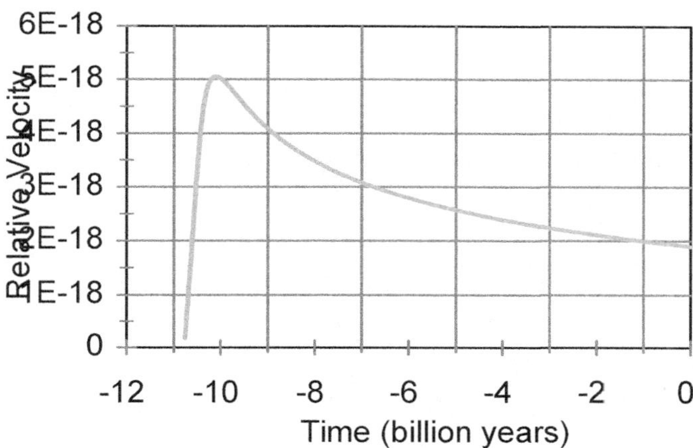

Figure 6.2 Cosmological expansion velocity.

The cosmological expansion velocity increased exponentially in the early stage of cosmic expansion and quickly reached a maximum after which it has been decreasing until it reached its current value.

The cosmic expansion has usually been presented as a deceleration and the deceleration predicted by the non-singular potential is shown in Figure 6.3.

Acceleration of the Universe

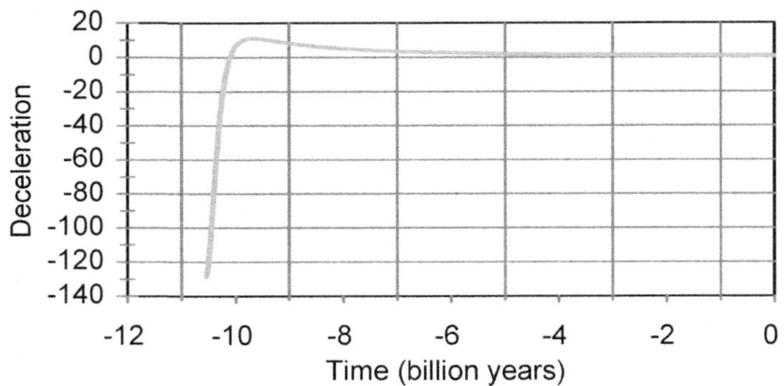

Figure 6.3 Cosmological expansion deceleration.

The picture of the cosmological evolution presented by the non-singular potential is very similar to the picture presented by the big bang model if inflation is included. The early times of cosmic evolution is dominated by exponential expansion starting from a point in time when the expansion velocity was zero. The difference between this picture and that presented by the big bang model is that here the size of the universe is not zero at the start time. Also, it is the force that the non-singular potential creates that causes the initial rapid expansion and later the deceleration of the expansion. All the features of the big bang model are here with the exception of the startup from a singularity.

One bit of experimental data that is said to support the big bang model needs to be discussed. The cosmic background radiation is taken to be support for the big bang model in that it is thought that from its initial extremely hot, small beginning the universe has expanded and cooled to the present time when it should be very cold. The measurement of the background radiation gave credence to this scenario. Indeed there was no other explanation for the background radiation at hand so it made sense to tie it to the cooling of the big bang. The discovery that the universe is basically five dimensional in space, time and matter gives rise to a five dimensional gauge field. Various aspects of a five dimensional gauge field of electric, magnetic and gravitational components will be presented in the next chapter. However, we should borrow one prediction of the five dimensional gauge fields to help distinguish between which theory the background radiation supports.

In the four dimensional gauge fields of electro-magnetism there are two concepts that become involved in this presentation. The first one is the concept of the energy density of the radiation. This is proportional to the sum of the squares of each field component. The other concept is that of radiation pressure. The radiation pressure is also proportional to the sum of the squares of the field components in four dimensions. This is not true in five dimensions. In five dimensional electromagnetogravitic fields the energy density is indeed proportional to the sum of the squares of all the field components. However, the radiation pressure is the sum of the squares of seven of the field components minus the sum of the squares of the remaining three components. The three components that must have their sums subtracted from the sum of the other components are the gravitational field components. There are experiments that will be discussed in the next chapter that are in support of this prediction.

The point applicable here is this; at zero radiation pressure there is a positive radiation energy density. It is difficult for me to understand how the universe can support a positive radiation pressure. Pressure on Earth requires something to push back when it pushes on something. I don't understand what that something is in the case of the universe as a whole. On the other hand, I can ask the question of whether there should be any radiation energy density in the absence of any radiation pressure. The answer in the five dimensional case is, yes, there must be a positive radiation energy density when the radiation pressure for non-zero field components is zero. The approximate magnitude of this radiation pressure is of the same order of magnitude as the experimentally measured cosmic background radiation.

Now there exist two potential predictions as to the existence of the background radiation. Another experiment is needed to separate the two predictions so as to tell which theory the data supports. There is an experiment that may be done in the lab that can verify the non-zero energy density for zero radiation pressure prediction. The test instrumentation is just now becoming accurate enough to simultaneously measure the energy density and radiation pressure to see if the five dimensional gauge field predictions are verified. If this prediction is verified than the cosmic background radiation is due to the five dimensional character of the universe and not because of a big bang start for the cosmological expansion.

Dark Matter

The tangential velocities of the stars in the arms of spiral galaxies do not act according to Newtonian gravity. Their velocities are slow enough that relativistic theories need not be used. Yet according to Newtonian gravity the tangential velocities of the stars way out in the outer reaches of the arms of spiral galaxies should drop off inversely proportional to the distance from the center of the galaxy, but they don't. In total disregard for Newton's gravitational force the tangential velocities of these stars refuse to drop off as they should. This has spurred no small amount of theoretical effort in an attempt to understand this data. In spite of this theoretical effort the community as a whole argues that the reason the tangential velocities do not drop off is that there must be some mass around each spiral galaxy that accounts for these velocities. This hypothetical mass cannot be seen in the normal sense of a radiating star so it is called dark matter.

When a fiend asked me to look into dark matter last year I applied the non-singular, time dependent gravitational field to the tangential velocities of the stars in the spiral galaxies' arms. I soon discovered that time dependence of the gravitational field correctly predicted the strange data for which dark matter had been needed. The primary reason for this comes from two aspects of the gravitational field. The first one is that the fastest speed at which a change in the gravitational field can travel is the speed of light. The second reason stems from the time dependence of the gravitational field which tells us that gravity is getting weaker all the time. This means that a star that is found at light years distance out in the arms of a spiral galaxy is responding to what the gravitational strength of the galaxy had been light years ago instead of what the gravitation is when the light from the distant star left the star or when it is received.

Further, it must be understood that it also takes light years for the data coming from the star out in the arm of the spiral galaxy to reach the Earth. This means that accurate predictions of what is happening based upon the light received must use the red shift predictions given above for the time dependent, non-singular potential instead of the old Hubble red shift predictions. Even so, the fact that the star in the arm of the spiral galaxy is responding to the greater gravitational strength of the mass of the galaxy at the time the gravitational signal left the center of the galaxy provides the major portion of the prediction that the velocities be different from Newtonian predictions.

Dark Energy

Emboldened by the fact that the time dependent gravitational field removed any need for dark matter to explain the tangential velocities of the stars in the arms of spiral galaxies, I proceeded to apply the time dependence to the dark energy issue. The need to hypothesize the existence of dark energy comes from more recent data using the assumption that the luminosity from type Ia supernova is a constant and comparing the distance predictions using the measured intensity of light from the supernova to the distance obtained due to the red shift data. This simplification of the issue does the study involved a disservice as the work done in all the supernova studies to date is phenomenally excellent work.

The analysis of the data uses Einstein's field equations and data from the red shift and magnitude of the light from supernova type Ia. Type Ia supernova were used because they are thought to be stars that collapse in on themselves at a point when they all have the same amount of mass. Having the same amount of mass when they collapse means they will emit light with the same luminosity. Then by identifying this type of supernova they should basically have a standard light source. If this is true the amount of light they receive from the distant supernova is a measure of its distance from Earth. If at the same time they find this type of supernova with large red shifts, meaning they are distant supernovas, the two distance measures should be able to tell the researchers a lot about the expansion of the universe. Their original objective was to measure the deceleration of the universe. Were they ever surprised by the data analysis.

What the data told the researchers was that cosmological expansion rate was speeding up rather than slowing down as everything argued it should. This finding initially caused the researchers alarm as it differed so much from expectations and then excitement as other researchers and data supported this initial contention. However, the only way in which they could see the data being explained within Einstein's theory was to resurrect Einstein's old

162

cosmological constant that provided a means of forcing outward expansion that their data showed was occurring. But such a cosmological constant argued that there was some energy in the universe that instead of attracting gravitating bodies it repulsed gravitational bodies. This energy was given the name of dark energy.

When I applied the time dependent gravity to the problem the most important result that impacted the conclusion of an accelerating rate of expansion was the influence of the time dependence on their assumption that the supernovas provided a standard candle. Some time ago a scientist by the name of Chandrasekhar studied collapsing stars. One type of star was held against collapse by the pressure of its electron gas. But once the mass of the star had increased to a certain point the gravitational force was able to overcome this pressure and the star began to collapse. For such stars the mass was always the same and this quantity was called the Chandrasekhar limiting mass. This star is the type Ia supernova chosen to be used as a standard candle because of the consistency of the quantity of mass at the time of collapse that caused the luminous explosion.

However, a time dependent gravitational field changes this analysis. Since gravity is getting weaker all the time it takes more mass now to have enough force to overcome the electron gas pressure than it did a long time ago. This means the Chandrasekhar limiting mass is time dependent and is growing larger all the time. The standard candle is getting brighter all the time. This means that older supernova will be dimmer because the Chandrasekhar limiting mass was less back when they exploded. Therefore, supernovas that are farther away, or older, would seem to be even farther away because of their dimmer light. This would give the impression that the universe had speeded up its expansion.

On the other hand, the time dependent gravitational field produces a time dependent Chandrasekhar limiting mass. This time dependent candle can be used in the analysis and the real cosmic expansion can still be extracted from the data. My study showed that there is no need to hypothesize the existence of dark matter to explain the data and no need for a cosmological constant.

Chapter 7 Electromagnetogravitic Fields

The five dimensional gauge field that gives rise to Einstein's general theory when mass is conserved has many surprises. First, there are five gauge potentials and ten gauge field components. The Weyl Gauge Principle allows us to derive the first order differential equations that tell how these ten field components interact. In Maxwell's electromagnetism there are five of these equations. Four of the equations are the basic Maxwell relations and the fifth equation is the conservation of electric charge. This equation is very similar to the conservation of mass that we have discussed before. When the full five dimensional gauge field is used there are eight equations including the charge conservation relation which now has an additional term added to it. It is these eight equations that will be discussed here

Force density vector

We can march down the same path as has been worked out in electromagnetism over the past century or so and derive many useful concepts. I did this and there are a lot things on my "to do" list of concepts to investigate as time permits. Not all of these concepts will be touched on here as I am not sure how to use them to add significantly to the story.

One of the first things that can be derived is the energy-momentum tensor. I actually used this in the derivation of Einstein's general theory of relativity. It is also the concept from which the force densities may be derived by looking at the five dimensional divergences. This is somewhat like looking at how the energy-momentum tensor varies with respect to the five variables in a manner similar to the procedure of obtaining the gauge potentials from the gauge function, but since the gauge function was a scalar and the energy-momentum tensor is a five by five tensor with 25 components it is a little more complicated.

The five dimensional divergence gives us five force density components and of these five components I really only wish to briefly discuss three. These three components correspond to the three forces with respect to space that constitute the forces with which we are familiar. These forces turn out to be the three components in space which contain the sum of the electrostatic, magnetic and gravitational forces. These are the force components that have been used in my research.

I have not gotten around to doing much with the other components, but within them there should be some enlightening information. The time component should have new information about how the energy flows with the added information about how the gravitational systems' energy behaves. The mass component should have the information that tells what forces mass to violate the conservation of mass relation. A study of this component stands to be very interesting.

Gauge field pressure

From the energy-momentum tensor we can get the expression for the radiation energy density. This tells us the density contained in any radiation. It may be worth while displaying this relation as we will be comparing it to another relation called radiation pressure. First, the energy density may be shown to be

$$\xi \equiv \frac{1}{8\pi} [\ \overline{E} \bullet \overline{E} + \overline{B} \bullet \overline{B} + \overline{V} \bullet \overline{V} + V_4^2 \ ] \ . \qquad (45)$$

which is the sum of the squares of all the gauge field components as I previously indicated. We can also derive the expression for the radiation pressure. One way of visualizing radiation pressure might be to think of all the photons running into a container wall and bouncing back.

166

The photons each have a little energy and this energy is flowing in the direction of the photons flight. Upon bouncing off the wall of the container the photons energy flow has changed its direction and this change of direction has imparted some reaction in the wall of the container. This is what gives the wall a stress that is due to radiation pressure.

The expression for this radiation pressure turns out to be given as

$$p = (\frac{1}{24\pi})[E^2 + B^2 + 3V_4^2 - V^2] \tag{46}$$

which has a minus sign on the last term. The first two terms are the electromagnetic components and are given as the sum of their squares just as in the radiation pressure. This is what gave us the interpretation that radiation pressure must be proportional to the radiation energy density. Actually this concept was put to a test when Nickels and Hull measured the radiation pressure and energy density of light. Their result was that the radiation pressure was a little less than the energy density, but their result was within their expected experimental error. Their error was of the order of 6%. In the early 1980's while I was still a military research associate at Los Alamos National Laboratory I sought, among the experts at the laboratory, a means of reducing this error. The instrumentation then used to measured the energy density of light was calibrated no better than 6%. This would not allow any improvement over the Nickels and Hull experiment and I could not get any funds to look into improving this accuracy even though a friend at the US Naval Academy recommended a means of using an interferometer to measure the energy density much more accurately.

If we take the Nickels and Hull data as accurate and use it to evaluate a parameter in the light they used that relates to the temperature of the light, this result can be used to calculate what temperature of light would cause the radiation pressure to go to zero. This temperature is very low, in the one digit range. However, this method of obtaining the value of the needed parameter has too much error to really specify the temperature very accurately. An improved accuracy in a simultaneous measurement of the radiation energy density and radiation pressure would allow a more accurate prediction of the temperature of zero radiation pressure. This temperature of radiation should then be compared to the temperature of the cosmological background radiation.

Five dimensional waves

My research into five dimensional waves has been an enlightening and frustrating study. It has been enlightening because of the new information it has produced and frustrating because of all the concepts I thought I knew because of my electrical engineering education that did not apply to the new waves. It has also been frustrating because of the difficulty of finding a solution to a set of wave equations that contains four wave equations that are inductively coupled.

I can spare the reader some of this frustration by stating that the four wave equations contain three vector wave equations and one scalar wave equation. A vector wave equation refers to a waving vector that has two values; a magnitude and direction. Examples of vector wave equations are the wave equations for the electric and magnetic fields. These wave equations are inductively coupled so that we only see a single wave called the electromagnetic wave. This wave has a component of the electric field and a component of the magnetic field traveling together. In classical four dimensional electromagnetic waves these field component are directed at 90 degrees to the direction the wave is traveling. Since the direction the field components point are perpendicular to the direction the wave is traveling these waves are called transverse waves.

In five dimensions where gravitational field components get inductively coupled to electromagnetic field components there also appears a transverse wave. However, this transverse wave has three components instead of just the two electric and magnetic components. The new additional component is a gravitational component which is parallel with but pointed in the opposite direction of the electric component. It is this gravitational component that causes the expression for the radiation pressure to be different from the expression for the radiation energy density.

Part of the frustration came from the fact that the five dimensional wave equations also have solutions that are independent of the transverse waves. This new set of waves also has three components; an electric and a gravitational component which are vector components and a scalar gravitational potential component. The scalar gravitational component is like the odd man out. I do not have a good feeling yet for this component. I just know that you may not have this set of waves without it. The electric and gravitational vector components of this wave set are pointed in the direction the wave is traveling. A wave with its vector components pointing in the direction of the propagation of the wave is usually called a longitudinal wave. To call this new wave set longitudinal waves would leave out any indication of the scalar component which must be there. So I have called these waves non-transverse waves to emphasize the fact that one component is not a vector.

You may have noticed that even though there is a gravitational wave equation it appears in combination with the other wave equations so there is no gravity wave independent of electromagnetic waves. The gravity wave does not exist by itself. There is a gravitational component in both the transverse and non-transverse waves. This may be a good reason the gravity wave experiments have come up negative.

Neutrinos

Earlier I presented the quantization of the electrostatic potential and the vector potential. The five dimensional gauge function has five potentials. The electrostatic potential is obtained when you wish to see how the gauge function varies with respect to time. There are three gauge potentials that contain the information of how the gauge function varies with respect to the three space variables. The last gauge potential tells how the gauge function varies with respect to mass. It is also this gauge potential that leads to the gravitational field.

If we look at the wave energy of this gauge potential just as we did for the photon, we find that we get a similar answer. The energy must be quantized and be proportional to the square of the quantum number and the frequency. That is it looks like

$$\varepsilon = N^2 h_n \nu. \tag{47}$$

I have used the subscript n to indicate that though the energy is proportional to the frequency and the square of the quantum number I cannot yet show that this constant of proportionality is Planck's constant.

This relation looks identical to that of the phat photon and, indeed, would be if the constant of proportionality turns out to be Planck's constant. So what might this wave/particle be? It must be associated with the non-transverse waves since the photon is associated with the transverse waves. If we look at the possible interaction between the electric field of the non-transverse wave with the electrons of the medium that the non-transverse wave may be traveling through we see that there is small chance of an interaction because the electric field is in direction of

170

the wave propagation. There are few particles that travel through materials without much interaction and one of these is the neutrino. Could our particle of non-transverse wave energy be the neutrino?

Chapter 8 Now Where Are We?

The flow chart at the end of Chapter 1 is probably the best way to guide a discussion of an overview of where we are so I will put it here also. Please refer to it as you may find it helpful and I will refer to it during my summary.

I put the fundamental laws in rectangular boxes with double end bars and restrictive assumptions in ovals to help distinguish them. I also highlighted the fundamental laws in red and the restrictive assumptions in yellow. The difference between them is significant and it is important to keep them separate. Fundamental laws are laws from which all phenomena of nature may be derived. They have no justification other than they describe nature. Restrictive assumptions are used to select from the totality of possible things that one may apply the fundamental laws to in order to either achieve a tractable statement of a problem or to study a restrictive subset of the totality described by the fundamental laws. Much of my research was aimed at determining if the fundamental laws could lead to the standard model of physics by restrictive assumptions. Therefore, there are several avenues of potential research that I did not follow so as to complete the objective I set for myself. Though I must admit my attention got diverted many times.

The statement of the First Law is the statement of conservation of energy as done in classical thermo-dynamics. This leads easily into three routes of inquiry. One route is that of assuming all mechanical forces are zero. This is the realm of classical thermodynamics without the intrusion of mechanics and is well presented in any number of texts and not really addressed here. This route is depicted on the far left of the flow chart. The second route is to assume the thermodynamic force is zero. This is the route to disclose the whole of mechanics without the intrusion of thermodynamics. This route leads to electromagnetism, quantum mechanics, special relativity, nuclear physics, and Newtonian mechanics. Down this path (far right on flow chart) those stable particles must have quantized electric charge and a non-singular electrostatic potential. This path also leads to forces that describe the weak and strong nuclear forces through behavior of the non-singular potential. It also leads to some new physics such as the phat photons. The third route is to assume both mechanical and thermodynamic forces are non-zero. This route leads to a five dimensional manifold of space-time-mass. It is down this route that conservation of mass leads to Einstein's general theory of relativity. This route also requires a time dependent gravitational field that predicts new expressions for red shifts, dark matter and dark energy.

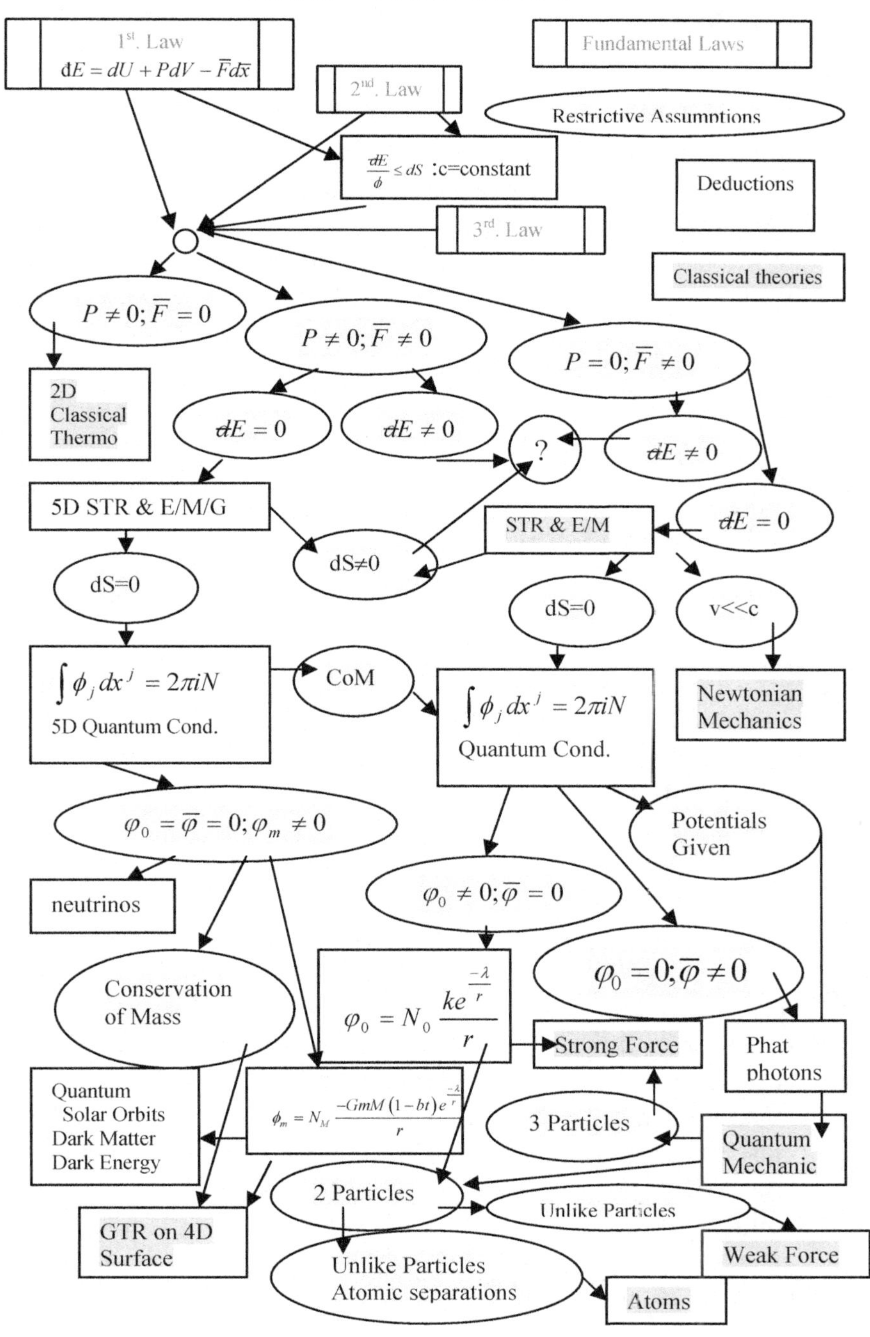
Fundamental Laws
Restrictive Assumptions
Deductions
Classical theories
1st. Law
dE = dU + PdV − F̄dx
2nd. Law
dE/φ ≤ dS :c=constant
3rd. Law
P ≠ 0; F̄ = 0
P ≠ 0; F̄ ≠ 0
P = 0; F̄ ≠ 0
2D Classical Thermo
dE = 0
dE ≠ 0
?
dE ≠ 0
dE = 0
5D STR & E/M/G
dS≠0
STR & E/M
dS=0
dS=0
v<<c
∫ φ_j dx^j = 2πiN
5D Quantum Cond.
CoM
∫ φ_j dx^j = 2πiN
Quantum Cond.
Newtonian Mechanics
φ_0 = φ̄ = 0; φ_m ≠ 0
Potentials Given
neutrinos
φ_0 ≠ 0; φ̄ = 0
Conservation of Mass
φ_0 = N_0 ke^(−λ/r)/r
φ_0 = 0; φ̄ ≠ 0
Strong Force
Phat photons
Quantum Solar Orbits Dark Matter Dark Energy
φ_m = N_M (−GmM(1−bt)e^(−λ/r))/r
3 Particles
Quantum Mechanic
2 Particles
Unlike Particles
GTR on 4D Surface
Unlike Particles Atomic separations
Atoms
Weak Force

Unification of the branches of physics

Not long after Einstein published his general theory of relativity he and others began a search for a way of unifying the forces of nature. At first they were just trying to unify the gravitational and electromagnetic forces as the weak and strong forces had not yet appeared. After the nuclear forces showed up researchers added them to the forces they were trying to unify. While I was in sympathy with their goal, I differed with them in at least two respects. First, I thought it rather odd that they tried to unify the nuclear forces with the electromagnetic forces without first addressing the implicit assumption that they were independent of the electromagnetic forces. Secondly, I felt that they really should be unifying the branches of physics rather than the forces. If they unified the branches of physics I felt the forces would take care of themselves.

Here we find that by starting with the thermodynamic laws it is natural to apply the restriction to isentropic systems when one wishes to find very stable systems. It came somewhat as a surprise that when this restriction was applied to particles a non-singular potential came popping out. This new potential form changed how we can look at the small (atomic), very small (nuclear) and the super small (sub nuclear). Plus, the potential does this while still holding the Maxwell equations to their classical form. It is the fact that when you add the discontinuous solutions to the set of continuous solutions to the Maxwell equations you get the non-singular potential. If all you are interested in are the continuous solutions you get the classical singular potential. This means that we now need to keep track of whether or not we wish to use only continuous solutions or do we wish to include the discontinuous solutions in the problem we seek a solution for. This is an interesting how do you do.

176

Another aspect of starting with the thermodynamic laws is that when we include all four force types we get a five dimensional manifold of space-time-matter in which the gravitational field is found to be, not only non-singular, but a function of time. The non-singularity of the gravitational field eliminates the singularity of a big bang yet it keeps an early exponential cosmological expansion of the universe. This time dependent gravitational field also predicts the effects in cosmology that are currently being used to argue the need for dark matter and dark energy. The tangential velocities of the stars in the far reaches of the arms of spiral galaxies are responding to gravitational field strength many lights years before their motion is measured and this gravitational field strength was then much greater than at the time of velocity measurement. In the case of dark energy the apparent accelerating expansion of the universe is the effect of assuming the type Ia supernova have always put out the same magnitude of light. The time dependent gravitational field means those older supernovas produced less light and, therefore, appear to be farther away than a standard candle would predict. So neither dark matter nor dark energy is needed to explain the data.

The point here is that the non-singular, time dependent gauge potentials help in understanding the nuclear and sub nuclear and the cosmological realms of the universe. This argues that a unification of the branches of physics has been achieved by adopting the classical laws of thermodynamics as the fundamental laws of nature. Not only is there no need to discuss different, virtually independent branches of physics, but all four forces are produced from one gauge function in five dimensions. Therefore, a unification of the forces was achieved at the same time the branches of physics were unified.

Several new experiments have been determined, some discussed and some not discussed, that can experimentally test different aspects of the predictions. For example, the phat photons prediction seems very testable with relatively small funding requirements. The inductive coupling and nuclear fusion are even now being tested. The results to date have all supported the predictions and these results should soon be published. The simultaneous measurement of radiation energy density and radiation pressure may cost a little more to do than testing the phat photon concept, but is doable with the right experimental setup.

New fundaments

Einstein was not afraid of thinking thoughts not previously conceived. Yet when he contributed so much to the beginnings of quantum mechanics, those who pursued quantum mechanics as a fundamental basis for physics felt they had lost a leader when Einstein steadfastly refused to follow their path. It is now possible to show how correct he was in maintaining his stand with the same rigorous logic that Einstein demanded of himself. There does indeed exist a simple set of fundamental postulates from which it has been shown that the basis of all the various branches of physics are but subsets of the totality of their description.

The starting point of this new line of thinking is so improbable as to be easily overlooked and yet it is the only foundation that has never been seen to offer predictions that differ from experience. This starting point is the laws of classical thermodynamics!

There are at least two reasons that classical thermodynamics would not be expected to provide such a foundation. First, thermodynamics, as currently studied, does not provide a description of motion like the mechanistic theories do. Secondly, texts teach, as Einstein believed, that classical thermodynamics might be obtained from statistical procedures applied to Newtonian mechanics.

178

The key insight needed to understand the fundamental nature of the laws of thermodynamics is to note that the first law is a Pfaff differential equation and to apply the second law of thermodynamics as Caratheódory did in 1909. Caratheódory's principle guarantees the existence of a property called entropy along with the energy statement of the first law. The form of these laws are such that they may be expressed, without preference, in any coordinate system and of any dimension, as Einstein stated should be required of a fundamental set of laws. Though the necessary, complimentary existence of energy and entropy appears to complicate any mechanistic description of nature, it is their simultaneous existence that provides a logical description.

Today, the concept of entropy is almost universally related to order or information. However, the concept demanded by the second law is best thought of as 'energy that becomes unavailable' as the thermal engineers have been known to call it. In this form, it is easier to connect the second law with the denial of perpetual motion. The more you do the greater the amount of energy that becomes unavailable. This becomes the entropy principle for isolated systems. For all other systems, it requires the minimum free energy principle. This provides variational principles that may be used to determine motion should a geometric metric also be given.

Now there is a surprising, but encouraging, immediate result from these laws. Using the same logic Caratheódory used to prove that the connection between energy (heat) and entropy is strictly a function of temperature and that this function leads to an absolute limit in temperature, these laws provide a function of velocity as the connecting function between energy and entropy for mechanical systems. More importantly, these laws provide a universal (i.e. independent of the type of force) limiting velocity for the same inertial systems that Einstein used in his special theory of relativity. There is now no need to assume that this as a fundamental postulate.

Bolstered by the fact that the laws seem to support the special theory of relativity by providing Einstein's postulate concerning the constancy of the speed of light, there is more reason to believe that mechanics may indeed come from these laws. Again following the lead of thermodynamics, in which there is a choice of the variables one may choose with which to write the second order differential equations that give the stability conditions, there is a choice provided for the description of mechanical systems stability. These differential equations are natural metrics for use in determining motion. Choosing the stability conditions in a manifold of space and entropy seems very unlike mechanics. Yet, when this metric is scaled using local time as the arc length, a rewarding requirement occurs when an isolated system is considered. These fundamental laws do not provide a variational principle in time. Therefore, the scaled metric must be solved for the element of entropy before the principle of increasing entropy may be applied. This gives equations of motion in a Riemannian manifold as Einstein used in his relativistic theories.

The two most remarkable features of this result are that the fundamental laws specify the geometry and that the laws require *two* metrics. The first metric is the relativistic metric whose arc length is the entropy playing the role of Einstein's proper time. The second metric is a somewhat similar relativistic metric with the energy as the arc length. This is the same requirement that occurs in thermodynamics where the change of entropy does not depend upon how the system changes. It depends only upon the end points while the heat depends upon how the change occurs. Here the distance between two points in the entropy manifold does not depend upon the path, but the distance between the same two points in the energy manifold does

180

depend upon the path. The laws require the energy metric to have a geometry that was first developed by Weyl when he proposed this geometry as a means to unify the electromagnetic and the gravitational fields in terms of the gauge fields that appeared within the geometry. Einstein argued that the path dependence of Weyl's geometric differed from experience with the result that Weyl's thoughts were abandoned so far as they went toward any unification.

In 1922, however, Schrödinger noticed that, should one require a unity scale in a Weyl space, only Bohr's quantized paths were allowed. Schrödinger went on to develop his wave equations of quantum mechanics in 1926. In 1927, London showed that the requirement of unity scale in a Weyl space could only be satisfied by paths that obeyed Schrödinger's wave equations. Further, London showed that Schrödinger's wave function was proportional to Weyl's scale factor. Weyl seized upon this result and raised London's result to the level of a principle, referred to it as Weyl's quantum principle. Weyl's quantum principle, and his display that the gauge potentials formed the scale factor in his geometry, led to the electromagnetic gauge fields. Providing the basis for all the subsequent gauge field work that Einstein referred to in 1940 and the work that has followed in the search for a description of the weak and the strong nuclear forces.

This historical review is of importance when very stable mechanical systems are sought. These would be represented by systems with constant entropy, called isentropic systems. Isentropic mechanical systems must have the unity scale of Weyl's quantum principle. Therefore, stable isentropic mechanical systems must satisfy quantum mechanics! Quantum mechanics is required by the fundamental laws only for the isentropic subset of all the mechanical systems in nature. Thus, quantum mechanics is required by these laws, but may never be considered by them to be fundamental to the description of all motions in nature. Furthermore, London's result that Schrödinger's wave function must be proportional to the Weyl scale factor, limits the statistical interpretation of quantum mechanics. Rather, London showed that the wave function contained information about the tendency of the scale factor to vary around unity. This requires the same mathematical operations as are currently used, yet with an entirely different interpretation.

One more result, from the fact that the entropy principle provides the variational principle for descriptions of motion, is one Einstein suggested was needed for any proper foundation of physics. The need for an explanation of inertia is satisfied by the fact that the entropy is an extensive property of the system requiring it to be proportional to the mass. This is the origin of the multiplicative mass term in the equations of motion.

The unsatisfactory separation between the theories of relativity and quantum mechanics, referred to by Einstein, is also hereby resolved. All isolated systems must obey relativity theory. Isentropic systems must also obey the laws of quantum mechanics.

Weyl showed that Maxwellian electromagnetism might be derived from his gauge potentials. The isentropic condition requires these potentials and the subsequent electromagnetic fields to be quantized. This requires the electrostatic potentials to be quantized in integer values as is seen by experiment. Existing theories do not require this quantization of electric charge. When the quantized electrostatic field is forced to satisfy the Maxwell equations, the dependence of the electrostatic potential upon space is determined to be a non-singular potential $[(1/r^2)\exp(-\lambda/r)]$ with the familiar $1/r^2$ long range dependence. The point, λ, at which the potential begins to deviate significantly from Coulomb's $1/r^2$ potential is different for different particles! This produces a violation of

182

Newton's action and reaction law for unlike particles that get very close to each other. Yet this has been shown to provide the basis of the nuclear force currently described as the weak force. Also, the equations of the Yang-Mills theory have been derived from this requirement for non-singular interactions between unlike particles that do not obey Newton's law of action and reaction. On the other hand, the non-singular potential requires that like particles change their character as they approach each other. This means that like particles, such as protons, that have a repulsive long-range interaction will have an attractive interaction as they are forced into the near proximity of each other. The equations for non-singular interactions between like particles have the SU(3) group characteristics.

How is gravitation required of the fundamental laws? Suppose one looks into the equations that describe a four dimensional hyper-surface that is embedded into a five dimensional Weyl manifold that have the four dimensions of space-time and an unknown, but physically real, fifth dimension. This is somewhat like asking for the description of a sphere, upon whose surface there are only two dimensions, that is embedded into a three dimensional space. The surface of the sphere will be curved even though the surrounding space need not be curved, but may have a Euclidean geometry. Without knowing what the fifth dimension might be, restricting it to be conserved similar to the statement of the conservation of mass provides a restriction that may quickly be explored. The equations resulting from the conservation of the fifth dimension are seen to be identical in form to those chosen by Einstein as his gravitational field equations in his general theory. Now the determination of the fifth dimension may be seen for the only physically real property that could give Einstein's equations is gravitating mass!

Substantiation for this conclusion may be had when one looks at the first law of thermodynamics and includes three spatial work terms plus the thermodynamic work term. The usual manner of writing the thermodynamic work term involves the specific volume. However, the reciprocal of the specific volume is the mass density. Therefore, the fundamental laws allow the mass as a fifth dimension. When this property is conserved, which is the only condition Einstein considered, there are at least three ways these laws describe gravitational phenomena. First, they may be described using the five dimensional gauge field descriptions, wherein the electromagnetic and gravitational fields form a single, inductively coupled, electromagnetogravitic field. Secondly, either of the two fundamental metrics for a surface embedded into a manifold may be used. The fundamental metric of the second type produces the Einstein equations. Further, Einstein's principle of the equivalence of inertial and gravitational mass is a further requirement of the fundamental laws and need not be made separately.

The fundamental laws require the quantization of gravitational phenomena for isentropic systems as well as a non-singular gravitational potential. The appearance of the non-singular gravitational potential changes the interpretation of black holes, the big bang, and red shifts of cosmological objects. Now the tie between gravitation and quantum mechanics has been established.

One last feature of these fundamental laws should be mentioned. It concerns Einstein's position that two separate theoretical descriptions of light, on the one hand as a particle and on the other hand a wave was intolerable. Electromagnetic waves follow from Weyl's gauge fields; that is, from the Maxwell equations. Isentropic propagation of electromagnetic energy must also satisfy Weyl's quantum condition and hence, must simultaneously satisfy the wave equations and be quantized. Further, the fundamental laws require that the quantized, isentropic propagation of electromagnetic energy must satisfy Plank's

184

blackbody radiation law. The wave and the particle nature of light are, therefore, both required by these fundamental laws.

Einstein stated that there appears to be two choices for a foundation for physics; statistical or deterministic. Here we see a foundation that is fundamentally deterministic. Non-isolated systems and systems with variable entropy must be deterministic while isentropic systems must be quantized and, therefore, may have a statistical nature even though the probabilistic interpretation of Schrödinger's waves was shown by London to be in error. Einstein's desire for a logically simple foundation for physics is also satisfied; for these laws have been shown to produce the foundations of each of the various branches of physics without yet coming upon a measured difference from experiment.

A return of Einstein's desired determinism to the foundations of theoretical physics, however, is insufficient justification to overturn decades of scientific thought. Even the pedagogic value of a more simple set of fundaments may not be adequate justification, as strong as it may be. A different explanation of already explained phenomena supports virtually no additional justification, especially if these explanations concern an understanding of cosmological development that does not affect everyday life. Rather, for some, adequate justification for a shift to a new foundation of theoretical physics, with its attendant relearning, rethinking and rewriting of physics, may come from new predictions of a practical improvement to human life.

One way in which new fundaments of theoretical physics may influence human life is in changing its view of reality. For example, the currently accepted physics has been used to establish our current view of reality to the point that it is commonplace to hear of a fundamental statistical origin for everything in the universe or a universally constant uncertainty in knowledge. Neither of these views of reality is supported by the new fundaments. All non-isolated systems and some isolated systems are fundamentally determinate. Isolated, stable (i.e. with no change in proper time) systems display quantum features and some of these justify using the statistical assumptions. This would certainly not support an evolution of things by chance as all systems that may have a statistical interpretation must be time invariant.

On the other hand, the new fundaments not only require that the unit of action (i.e. uncertainty) depends upon the system under consideration, (e.g. an atom, a nucleus, or a solar orbit), will have a different quantum unit of action, but, further, of the four universally accepted fundamental values of electric charge, Planck's constant, the gravitational constant and Hubble's constant, only two are truly fundamental while the other two may be derived. This displays an unprecedented unification as well as a simplification.

The brief discussion above mentions numerous new concepts that stem from the new basis. The equations underlying these concepts provide several experiments that may test these concepts, yet proof of the new theoretical basis does not, of itself, provide a prediction of improvement to man's life. They have been used to illuminate a few areas in which appropriate engineering solutions may achieve significant advancement.

The inductive coupling between the electric and the magnetic fields, predicted by Maxwell's equations, has led to uncountable products involving the interplay between these fields. In particular, we have learned to electrically make magnetic fields or the use magnetic fields to make electricity. The inductive coupling between the electromagnetic and gravitational fields predicted by a five dimensional gauge field provides an extremely fertile ground for engineers.

186

The five dimensional wave equations require the transverse waves to consist of an electric, a magnetic and a gravitational component rather than just the electric and magnetic components. This leads to the prediction that the electromagnetic energy density be non-zero when the radiation pressure vanishes. This suggests two things. First, since it is difficult to imagine the universe supporting a non-zero radiation pressure, then there must be a non-zero electromagnetic energy throughout the universe as is being measured. Secondly, this provides a new view of the zero point vacuum energy that may be more receptive to an engineering approach to mining it.

Another way new fundaments of theoretical physics may have an impact upon humans is to provide new logical basis upon which to look at our universe. This can lead to new understandings of known phenomena or to exciting predictions of new physics. For example, the study of the energy radiating from a blackbody led Planck to the first assumption of quanta and the first successful equation of quantum mechanics. What of the study of the blackbody itself? Obviously, a system radiating energy should not be considered to be isolated. Non-isolated systems have not been discussed above where the concentration was on isolated systems. An electron under the accelerating influence of a force that radiates energy is an example of a non-isolated system. So is a blackbody. The new fundaments of theoretical physics provides a variational principle in the minimum free energy principle and this principle should provide the equations of motion for these systems

Modifications to current models/interpretations

Many current models and interpretations of physics will need to be changed in light of the new foundations of physics contained in the preceding chapters of this book.

Conservation of energy

Many concepts and equations have been given the interpretation and name of conservation of energy. In many cases the interpretations were slightly misleading, if not wrong, and the equations were only approximations to the real conservation of energy. These cases were almost all restrictions made to get a better understanding of a problem or a better route to a solution. Most of them missed the mark of conservation of energy without actually stating how they missed the mark.

An example may be had by looking at the difference between the conservation of energy as given by the First Law (see Equation (3)) and the conservation of energy statement that comes from Newtonian mechanics. The difference stems from the fact that in the First Law the mechanical forces must depend upon the position and velocity. The reason they must depend upon both the position and velocity is that the First Law is a path dependent statement of the conservation of energy. However, the statement of conservation of energy in Newtonian mechanics involves forces that depend only upon the position. How do they compare?

The First Law has forces given in the form of Equation (7). This force depends upon both the position and the velocity. Choosing the First Law to be the true, real, or master statement of the conservation of energy this must be the form of the force to satisfy conservation of energy. However, suppose we wish to consider the low velocity limit of the force given in Equation (7) which is only a function of position. We could then write Equation (7) as

$$ma = F \approx F(x). \tag{48}$$

This is just the Newtonian force that is strictly a function of position only. It leads to the fact that the mechanical energy of the system is the sum of kinetic energy (energy due to

188

motion) and potential energy (energy due to position) must be a constant. This is the first statement of the concept of conservation of energy that the student of mechanics runs into in their studies. It is taught as THE conservation of energy. We now see that THE conservation of energy is given by the First Law and the old familiar mechanical conservation of energy is the low velocity limit of the First Law. We must keep in mind that when using this low velocity limit the path dependence of the First Law is lost.

Conceptually and physically the loss of the path dependence makes a significant difference between the First Law and its low velocity approximation. One way of seeing this is to note that while it is easy to see that the low velocity approximation quickly follows from the First Law for slow moving things. However, one would be hard pressed to guess or determine the First Law given only the low velocity approximation. The low velocity approximation is a subset of the First Law, but is not its equal.

We may next consider the First Law with the velocity dependent forces and learn that these will form a relativistic statement of conservation that is referred to as the conservation of the four-momentum. This is the statement that the sum of the squares of the three components of the momentum subtracted from the system's energy divided by the speed of light must equal the square of the rest energy given by mc^2 divided by the speed of light. This is a true statement of conservation of energy as given by the First Law. Or is it?

Remember that the four dimensional relativistic physics of relativity comes from the First Law when mass is conserved. So we see that even the relativistic statement of conservation of energy, while closer to the First Law than the Newtonian conservation of energy, is still but a subset of the First Law as relativity was obtained by assuming no dependence upon the fifth dimension.

Unit of action

Quantum mechanics rests upon the quantization of the unit of action. In atomic physics this shows up as the quantization of the angular momentum in the atom. In general quantum mechanics it shows up in the quantum Poisson brackets. The unit of action is associated with Planck's constant. The uncertainty principle is also related to the quantum Poisson brackets and Planck's constant. This has led to the interpretation that Heisenberg's uncertainty relations involve a constant, namely Planck's constant. However, we have seen in the preceding chapters that the Poisson brackets are formed through the process of differentiation which involves geometry and that the unit of action depends upon the gauge function. This dependence of the unit of action upon the gauge function is true independent of any vector curvature that may exist.

This dependence of the unit of action upon the gauge function changes the interpretations of a Heisenberg's uncertainty principle that depends upon a universal constant. This change in interpretation plays a big role in the nuclear model contained in the preceding chapters as compared to the standard model. Mainly this mean uncertainty is not constant.

Big bang

The standard model of cosmology states that the universe started to expand from nothing at a singular point. One of the supporting pieces of experimental evidence is the existence of the cosmic background radiation. Herein the non-singular gravitational potential denies that the universe started from a point. Indeed when one looks backward in the cosmic development there is small, but finite size of the universe where the expansion velocity was zero. The universe then expanded exponentially from this small, finite radius. There was no singularity as in the big bang.

We found previously that the cosmic background radiation does not necessarily come from the cooling of the big bang. Though undoubtedly the universe was much hotter when it started at the small initial radius, the primary contributor to the presence of a cosmic radiation is the radiation energy required at the zero pressure boundary condition for the universe. That is, the cosmic background radiation is due to the five dimensionality of the universe and the fact that there can be no pressure supported by a cosmic vacuum.

Black holes

There are two types of objects that might be given the name of black holes. The name comes from the notion that a black hole is an object that is so massive that light cannot escape its gravitational field. However, black hole is also the name given to objects that are massive and small enough to meet the singularity in the Schwarzschild solution of Einstein's gravitational field equations. These two definitions of a black hole may not be identical.

The non-singular gravitational potential denies that there are gravitational singularities in the universe. This argues that there are no black holes that are gravitational singularities. I am not so sure that this rules out objects that can keep light from escaping. I have not yet investigated the ability of very dense gravitational bodies to pull light toward them to see if there exists any condition of mass that can prevent light from escaping such an object.

New predictions

In the preceding there are two types of new predictions. There are new predictions that pertain to known phenomena and there are predictions of entirely new physical phenomena. The following is a partial list of predictions that have been discussed in the preceding chapters.

Earth's magnetic moment

The Earth has been known to have a magnetic moment for several hundred years. It is a phenomenon that sailors have put to good use. Yet it is not predicted by any fundamental theory. It is explained by rotating currents within the Earth, but what drives the currents still need an explanation.

Above we saw that the inductive coupling between the electromagnetic and the gravitational fields requires that a spinning, electrically neutral, gravitational body must have a magnetic moment. Not only must the earth have a magnetic moment, but the value predicted using the charge-to-mass ratio is very close to the experimentally measured value.

Gravitational Rotor

The gravitational rotor was a device attempting to find a means of electromagnetically creating a gravitational field. The inductive coupling between the electromagnetic and the gravitational fields means that one can be used to create the other. It remains to find one of the many ways in which this may be done. The coupling is a weak one so whatever means is found it may require large electromagnetic values to create useful gravitational fields.

It may be true that we know more about how to manipulate electromagnetic fields than we know about manipulating gravitational fields. However, gravitating bodies in motion create electromagnetic effects that can be measured. The first example of this is the Earth's magnetic field. Another example of a gravitational body in motion is one that is falling in response to the pull of the Earth's gravitational attraction. Northrop Grumman has hired one researcher to investigate the measurement of electric effects of such gravitational bodies in motion. The results of this investigation has not yet been completed nor published, yet informal communication reveals the measured effects are in accordance with the predictions.

192

Cosmic background radiation

The cosmic background was measured several decades ago. It is thought to be the remnant of the hot, big bang origin of the universe. Earlier we saw that the zero pressure boundary condition in a five dimensional universe requires there to be a non-zero radiation energy density. This is the prediction of the cosmic background radiation. I have not yet gotten an accurate measurement from an experiment such as the simultaneous measurement or radiation energy density and radiation pressure in order to make a prediction of the frequency of the background radiation to compare with experiment. However, the rough number obtained is in the ballpark of the measured values.

Nuclear masses

The standard model has developed a semi-empirical mass formula with which to calculate the mass of the various nuclei. This model has been developed using different models of the nuclei, without the benefit of a functional form for the nuclear forces. The non-singular electrostatic potential was used herein to generate solutions to the energy equations for nuclei. The exponential character of the equations makes the determination of a mass number difficult, so approximations were made. The resulting predictions of the masses of the low mass nuclei were better than the predictions of the semi-empirical mass formula.

Neutron

In the standard model the neutron consists of three quarks, but no description of how these quarks hold the neutron together is possible. Above the non-singular electrostatic potential was used to predict the radius of a proton in orbit around an electron forming a neutron. This solution satisfies all the spin and energy requirements and also predicts the half life of the neutron that is not possible to predict in the standard model.

Deuterium

The deuterium nucleus was shown to be two protons in orbit around an electron.

Helium

The helium nucleus was found to be four protons in orbit around two electrons.

Fusion

It was shown that two deuterium nuclei could be caused to fuse into a helium nucleus with much less energy than required by the standard model. Plus, in this case the two deuterium nuclei preferentially fused into helium without neutron radiation or other radioactivity.

Phat photons

The isentropic requirement for radiation led to the prediction of photons with quantified energy where there were photons with greater energy at a given frequency than Einstein predicted. These photons were called phat photons. These particles were the carriers of the transverse wave energy.

Neutrinos

The same isentropic requirement that led to phat photons for the transverse wave also leads to the prediction of quantified energy of a particle carrying the non-transverse wave energy. This particle does not readily interact with material and, therefore, appears to be a neutrino. A form of communication system was developed that used field-field antennas to generate and receive these neutrinos.

Dark matter/dark energy

Dark matter has been hypothesized to account for the tangential velocities of stars in the outer reaches of the arms of spiral galaxies that do not obey Newtonian physics. Dark energy has been hypothesized to explain the appearance of older supernova being dimmer than current theories predict. Neither of these hypotheses is needed as the time dependent, non-singular gravitational field explains both sets of data without the need of a new hypothesis.

The tangential velocities in the far reaches of the arms of spiral galaxies do not obey Newtonian gravity predictions because they are responding to the strength of the gravitational field of the galaxy as it was light years earlier. Since the gravitational field is getting weaker with time the gravitational field of the galaxy was much stronger light years before the data was taken. When the gravitational signal left the center of the galaxy the field strength was much greater than it becomes later. This means the tangential velocities of the stars in the outer reaches of the arms of the spiral galaxies will not be reduced as the time independent Newtonian gravity would predict. Rather, they will follow the predictions of the time dependent gravity that causes the tangential velocities to remain greater than Newtonian gravity would predict.

The use of type Ia supernova as standard candles to compare the distance of these supernovas with their red shift data has resulted in older supernovas appearing dimmer than they should be given a decelerating expansion of the universe. Thus, using Einstein's gravitational field equations and a time independent gravitational field the data argues that the universe's expansion is accelerating and dark energy has been hypothesized to explain this accelerating expansion. However, the time dependent gravitational field that causes the gravitational field strength to weaken in time causes the Chandrasekhar limiting mass to change with time and, therefore, is not a good candidate as a standard candle that does not change with time. Rather, since the Chandrasekhar limiting mass is the mass required to overcome the electron gas pressure and cause the star to collapse and become a supernova a weakening of the gravitational field strength means that more mass is needed cause newer supernovas than for older supernovas. This means newer supernovas shine brighter than old ones. This is the primary explanation of the data without the need for dark energy.

Is the End in Sight?

There have been a lot of books and articles written about a Theory of Every Thing and the ultimate unification theory. Many times through the years I have heard people refer to my work as seeking either the grand unification or a theory of everything. I do not see it that way. Early on in this research I thought that unification of the various branches of physics was needed and such unification of the branches would necessarily unify the forces. I see now that while a unification of the branches of physics has been achieved this may be no closer to a theory of everything than we were previously as there are several avenues of research that have been opened up in this study that needs to explored. Every time I used a restrictive assumption to narrow down the research so that it could be aimed at the currently accepted theories this act of restricting the view leaves a realm of physics that needs more research.

For example, one early restriction made was to restrict our attention to only isolated systems. These are systems for which no energy is being exchanged between the system and its surroundings. In classical thermodynamics these are systems that are thermally isolated. In mechanical systems these would be systems that do not exchange energy with their surroundings during their motion. An example of such motion is the motion of an electron around a nucleus in an atom. This may not be the best example, since the motion of the electron in an atom also has the additional restrictive assumption of being isentropic motion. The isentropic assumption forces the electron to obey quantum mechanics. An electron that is being accelerated by a force and is also

196

radiating energy away may represent an example of an electron that is non-isolated and non-isentropic. Such motion would require an analysis using both the entropy and the energy manifolds where the isolated system needs only the entropy manifold. Another example of a non-isolated, non-isentropic motion may be the motion of an electron going from a higher energy orbit within the atom to a lower energy orbit while emitting a photon. This is a phenomena that cannot be reached by quantum mechanics and, hence, the argument that if quantum mechanics can't provide the answer you shouldn't ask the question. Here there should be an answer given that would provide additional knowledge to the motion of an electron between stable orbits.

The prediction that the electromagnetic and the gravitational fields are inductively coupled introduces an entirely new field of research the bounds of which would be hard to estimate. For example, such a coupling would predict a means of electromagnetically creating a gravitational force. It also means that one should be able to electromagnetically detect gravitating mass. Of course one means of electromagnetically detecting gravitating mass would be to look for a magnetic moment such as the Earth displays. However, one may also expect a mass in motion to leave an electromagnetic trail behind it. This prediction is even now being pursued. To date all the data of this avenue of research has confirmed the predation, but is being thoroughly checked before publication.

The new nuclear model that argues that nuclei are formed by the heavier protons orbiting around the lighter electrons opens up a new field of nuclear chemistry that is more like atomic chemistry. An investigation of allowed proton orbits may provide many new insights into nuclear physics. Also, the notion that a proton may be composed of two positrons in orbit around an electron adds an entirely new realm of research that might be called sub nuclear physics.

The time dependent, non-singular gravitational field opens up new vistas in astrophysics and cosmology. These fields could provide new and surprising information about our universe.

This is a very limited discussion of the possible new avenues of research that should support my contention that instead of approaching a theory of everything, the above, while perhaps a step in that direction, is a long way from getting close to such a goal. I see no need for the new student in physics to worry about there being nothing new left for them to study.

As a teaser one might ask if there is anything in nature that is independent of space, time and matter. I asked that question and surprised myself with the answer that comes from the above logic. The answer will be presented later.

The end is nowhere is sight.

Bibliography
1. J. D. Garcia and J. E. Mack, J. Opt. Soc. Am. **55**, 654 (1965).
2. C. E. Moore, NSRDS-NBS **3**, Sect. 6 (1972).
3. Williams, P.E., 1976, "On a Possible Formulation of Particle Dynamics in Terms of Thermo- dynamic Conceptualizations and the Role of Entropy in it," thesis, U.S. Naval Postgraduate School.
4. Williams, P.E., 1977, "The Principles of the Dynamic Theory," Research Report EW-77-4, U.S. Naval Academy.
5. Williams, P.E., 1980, "The Dynamic Theory: A New View of Space, Time, and Matter," Los Alamos National Lab report LA-8370-MS, December.
6. Williams, P.E., 1981, "The Dynamic Theory: Some Shockwave and Energy Implications," Los Alamos National Lab report LA-8402-MS, February.
7. Williams, P.E., 1981, "The Arrow of Time in the Dynamic Theory," Los Alamos National Lab report LA-8690-MS, February.
8. Williams, P.E., 1983, "The Nuclear Model from the Dynamic Theory," presented at the Third Annual Southwest Theoretical Physics Conference, at University of Colorado, Boulder, CO, March.
9. Williams, P.E., 1983, "The Possible Unifying Effect of the Dynamic Theory," Los Alamos National Lab report LA-9623-MS, May.
10. Williams, P.E., 1983, "The Potential of the Dynamic Theory in Modeling Electro-magnetic Effects on Biological Systems," presented at the Conference on "Non-Ionizing Effects of Electromagnetic Radiation on Biological Systems," at Veterans Hospital, Loma Linda, CA, May.

11. Williams, P.E., 1989, "Space, Time and Mass," Center for Explosives Technology Research Report A-07-89, May.

12. Rynne, Timothy M. and Dering, John P., "Experimental Investigation of an Electromagnetic-Gravitational Interaction," the Electric Spacecraft Journal, 1994.

13. Williams, P.E., "Quantum Measurement, Gravitation, and Locality in the Dynamic Theory," Symposium on Causality and Locality in Modern Physics and Astronomy: Open Questions and Possible Solutions, York University, North York, Canada, August 25-29, 1997.

14. Williams, P.E., "Thermodynamic Basis for the Constancy of the Speed of Light," Modern Physics Letters A, Vol.12, No. 35 (1997) 2725-2738.

15. Williams, P.E., "Using the Hubble Telescope to Determine the Split of a Cosmological Object's Redshift in its Gravitational and Distance Parts," Apeiron , Vol. 8, No. 2 (April 2001) http://redshift.vif.com/JournalFiles/V08NO2PDF/V08N2WIL.pdf

16. Williams, P.E., "Mechanical Entropy and its Implications," *Entropy* **2001**, *3*, 76-115. http://www.mdpi.org/entropy/list01.htm#new

17. Williams, P. E., "Energy and Entropy as the Fundaments of Theoretical Physics," *Entropy* **2002**, *4*, 128-141. http://www.mdpi.org/entropy/htm/e4040128.htm

18. Williams, P. E., "Alternate Communications for Space Travel," Space Technology and Applications International Forum (STAIF-2007), Albuquerque, NM, February 11-15, 2007.

19. Williams, P. E., "Compact Reactor," Space Technology and Applications International Forum

(STAIF-2007), Albuquerque, NM, February 11-15, 2007.
20. Williams, P. E., "New Time Dependent Gravity Displays Dark Matter and Dark Energy Effects," Apeiron Vol. 15, No. 3 (July 2008) http://redshift.vif.com/JournalFiles/V15NO3PDF/V15N3WIL.pdf

BIBLIOGRAPHY

Index

Printed in Dunstable, United Kingdom

85047120R00127